How to use this book

Welcome to Twenty First Century Science. This book has been specially writte a partnership between OCR, the University of York Science Education Group, the Nuffield Foundation Curriculum Programme, and Oxford University Press.

On this page you can see the type of page you will find in this book, and the features that will help you during your course.

Everything in the book is designed to provide you with the support you need to help you achieve your best.

Main Pages

Title: This tells you what this page is about and what you will learn about.

Main content: This is the important material that you need to know. The small chunks of information help you to learn at your own pace and make sure that you are well prepared for the certificate at the end of your course.

Main Pages

Diagrams: Lots of useful diagrams and images help you to visualise what you are learning and guide you through scientific concepts.

Questions: Use these questions to see if you've understood the topic.

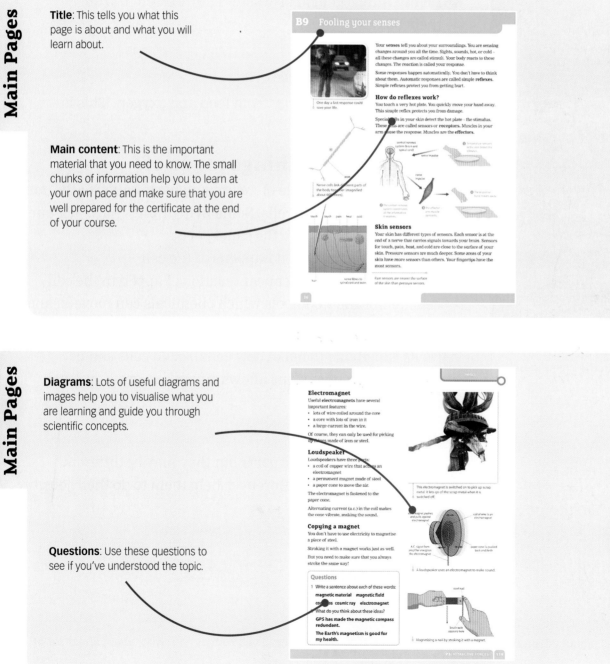

Being alive

Look at the photo of the greenhouse. The woman and the plants are living things. All living things have to do the same things to stay alive. These are called the seven processes of life.

1 Living things move.

2 Living things sense changes around them.

3 Living things need nutrition (food).

4 Living things get rid of waste. They excrete it from their bodies.

5 Living things grow.

6 Living things reproduce.

7 Living things get energy from food and oxygen. This is called respiration.

What are living things made of?

All living things are made of very small parts called **cells**. You are made of millions of cells. Some living things have just one cell. Animal cells all have the same parts:

• nucleus – controls what happens in the cell

• cytoplasm – where chemical reactions happen in the cell

• cell membrane – controls which chemicals can move in and out of the cell.

All cells can divide to make new cells, so old cells can be replaced by new ones. This allows animals to grow or repair themselves.

Specialised cells

Cells have the same basic parts but they may be different shapes or sizes. These differences help them to do their jobs.

What's alive in the greenhouse?

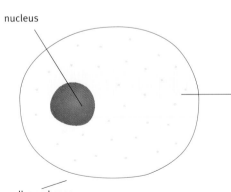

An animal cell (about 20 μm across).

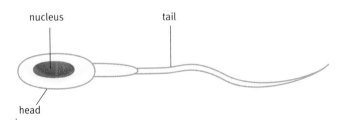

Sperm cells have tails so they can swim (about 60 μm long).

Cells work together

Cells work together to keep you alive. Cells are grouped together into **tissues**. Each tissue contains just one kind of cell. Tissues join up to form **organs**.

The stomach is an organ. It has muscle tissue in the wall. The stomach lining has tissue that makes digestive juices.

Organs work together in an **organ system**. Different organ systems do different jobs in your body:

- the circulatory system – keeps blood flowing to all parts of your body
- the respiratory system – gets oxygen into your blood
- the nervous system – links your brain to other parts of your body
- the digestive system – breaks down food so it can get into your blood.

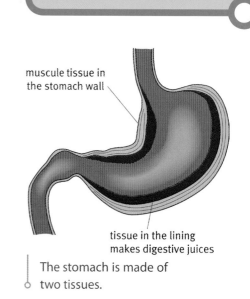

muscule tissue in the stomach wall

tissue in the lining makes digestive juices

The stomach is made of two tissues.

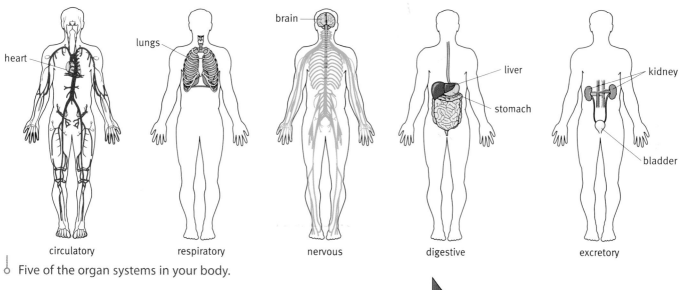

heart

lungs

brain

liver

stomach

kidney

bladder

| circulatory | respiratory | nervous | digestive | excretory |

Five of the organ systems in your body.

Energy for life

Each type of cell has its own special job to do but a cell can only do its job if it has energy. Cells get energy from food like glucose. Glucose and oxygen react together in the cytoplasm. This releases energy from the glucose. The reaction is called respiration. The waste products are carbon dioxide and water. These leave the cell through the cell membrane.

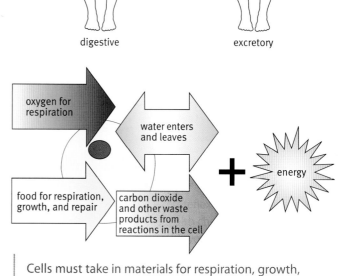

oxygen for respiration

water enters and leaves

food for respiration, growth, and repair

carbon dioxide and other waste products from reactions in the cell

energy

Cells must take in materials for respiration, growth, and repair. They must also get rid of waste. Respiration releases energy from glucose.

Exercise

Matt goes jogging. His leg muscles contract to push him forward. This means that his muscles need more energy. The cells need more glucose and oxygen for respiration. Matt's heart beats faster moving his blood around faster. Blood delivers more glucose and oxygen to the cells.

Matt's breathing rate also gets faster. His lungs work hard to get fresh oxygen into his blood. At the same time they get rid of carbon dioxide waste from his blood.

Matt stops running. His pulse rate and breathing rate go back to normal in a few minutes. This shows that he is quite fit.

Warming up

Matt always warms up before he starts running. Afterwards he walks slowly for a few minutes to cool down. This helps to stop his muscles getting injured. Matt tries to make sure he eats a healthy diet so that his muscles get enough energy.

Important organs

You can die if one of your organs stops working.

Cheryl's kidneys have stopped working. They are not getting rid of waste from her blood. To stay alive she has to spend two days each week connected to a dialysis machine.

When you run your pulse rate gets faster.

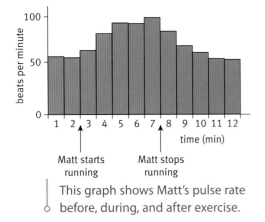

This graph shows Matt's pulse rate before, during, and after exercise.

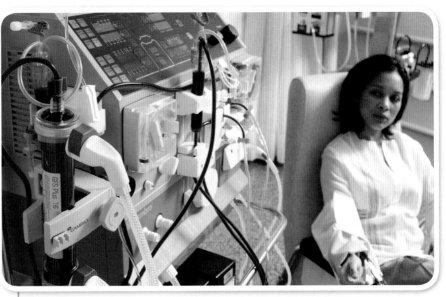

The dialysis machine removes waste from Cheryl's blood.

Kidney transplant

Dave has a car accident. Sadly he dies in hospital. Dave was carrying a donor card. The card tells doctors that he wanted his organs to help other people when he dies. The doctors remove some of Dave's organs, including his kidneys.

The kidneys start to decay as soon as they are taken out of Dave's body. They are kept next to ice to cool them down.

Cooling slows down decay. The kidneys must not be frozen. If ice forms inside cells they are damaged. Then the cells can't work properly.

Dave's kidneys are a good match for Cheryl. Doctors remove one of her faulty kidneys and replace it with one of Dave's. This is a kidney **transplant**.

After the operation, Cheryl is given drugs to prevent her body from rejecting the transplanted organ. Her new kidney works well. Cheryl stops using dialysis and leads a normal life.

NHS Organ Donor Register

donorcard

I want to help others to live in the event of my death

I request that after my death
A. any part of my body be used for the treatment of others ☐, or
B. my kidneys ☐ corneas ☐ heart ☐ lungs ☐ liver ☐ pancreas ☐ be used for transplantation.

Signature _____ Date _____

Full name _____
(BLOCK CAPITALS)
In the event of my death, if possible contact:

Name _____ Tel. _____

Remember to tell someone close to you that you want to be an organ donor. We'll need their agreement if the time ever comes.

Doctors know you want to donate your organs if you carry a donor card.

Doctors keep a donated organ cool so that it is in good condition when they transplant it.

Questions

1 Write a sentence about each of these words:

   cell   tissue   organ

   organ system

   transplant

2 What do you think of these ideas?

   Only living things can make copies of themselves.

   Everyone should carry an organ donor card.

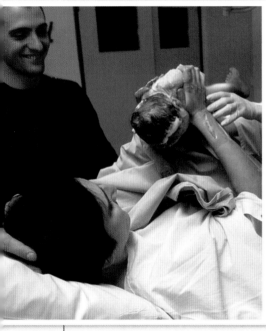

A newborn baby boy.

This baby has been growing inside his mother for nine months. In this time he has gone through amazing changes. You went through these same changes.

How does it begin?

All the cells in your body came from just one cell. This was a fertilised egg cell. Fertilisation happens when a **sperm** cell joins with an **egg** cell.

A sperm is a special cell made by a man in his testes. Each man makes millions of sperm. The sperm has a tail so that it can swim towards the egg cell.

A woman's ovaries release one egg cell every month. The egg cell passes down the oviduct towards the womb. If the egg cell meets a sperm along the way, it may be fertilised.

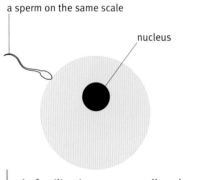

At fertilisation an egg cell and sperm join up.

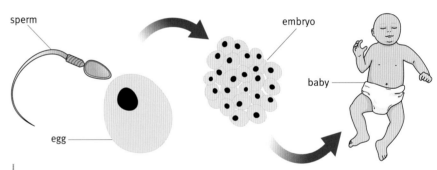

Birth takes place about 40 weeks after a sperm fertilises an egg.

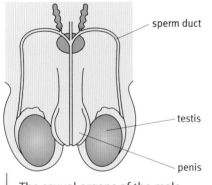

The sexual organs of the male.

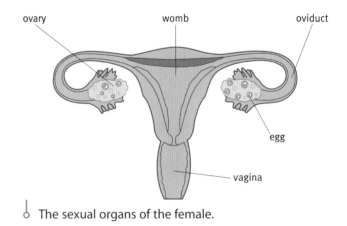

The sexual organs of the female.

Fertilisation

For fertilisation to happen sperm must get inside the woman's body. This happens during sexual intercourse.

- The male puts his penis into the female's vagina.
- Sperm move along the sperm duct and out into the vagina.
- The sperm swim through the womb towards the egg cell.
- The sperm meet up with the egg cell in the oviduct.
- One sperm wriggles into the egg cell and fertilises it.

After fertilisation

After an egg is fertilised it starts to divide. It is called an embryo. The **embryo** settles in the thick lining of the mother's womb. The embryo cells keep dividing. After about two months the embryo becomes a **fetus**. Now it has enough cells to make organs.

Sometimes the embryo splits into two at a very early stage. The two groups of cells develop into separate fetuses. They grow into identical twins.

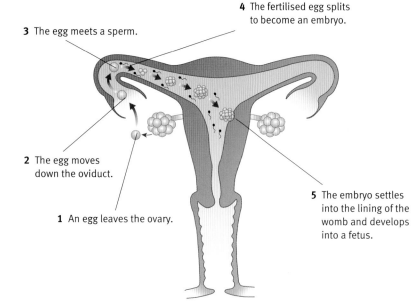

4 The fertilised egg splits to become an embryo.

3 The egg meets a sperm.

2 The egg moves down the oviduct.

1 An egg leaves the ovary.

5 The embryo settles into the lining of the womb and develops into a fetus.

Sometimes two or more eggs are fertilised at the same time. Each can grow into a separate fetus. They grow into non-identical twins.

Baby

The baby grows inside a bag of water. This stops microbes getting to it. The water also cushions the baby when the mother moves. Blood from the baby flows along the **cord**. This blood carries wastes away from the baby. The blood flows into the **placenta**. At the placenta:

- wastes pass from the baby's blood into the mother's blood
- food and oxygen pass from the mother's blood into the baby's blood.

The cord takes the baby's blood back again.

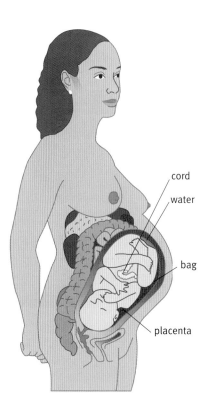

cord

water

bag

placenta

Pregnant

How does a woman know if she is pregnant?

I've missed a period. Am I pregnant?

Use a pregnancy test on your urine – that will tell you.

It is important to take care of the woman and her baby.

They checked my blood pressure, weight, and height at the clinic.

You ought to give up smoking and drinking – it's better for the baby.

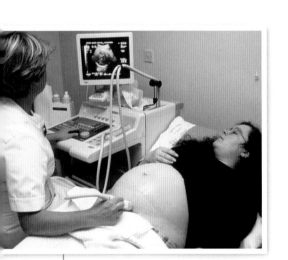

Ultrasound scans show how the baby is developing.

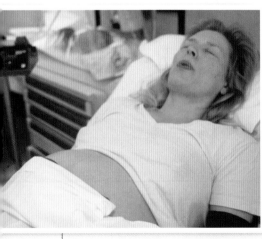

Muscles in the womb squeeze the baby out.

Birth

It takes about nine months for the baby to grow fully. Then it is time for him or her to be born. The mother goes into labour.

To start with, the muscles in the womb squeeze on the baby. This happens every few minutes, and it can hurt the mother. The womb squeezes harder and harder, and more and more often. This hurts more.

Suddenly the bag breaks and the water escapes. This is messy.

Every time the womb squeezes its opening gets bigger. When the opening is big enough, the squeezing of the womb pushes the baby out. This can really hurt.

Once the baby is out, it starts to breathe. The baby doesn't need the cord any more, so it can safely be cut.

Finally, the placenta is squeezed out of the womb. This is called the afterbirth.

The mother's periods start again soon after the birth.

Pain relief

Giving birth can be painful. There are some ways of making it easier:
- breathe in drugs to reduce the pain
- have an epidural, an injection of anaesthetic into the lower spine
- learn relaxation and controlled breathing to ease the pain.

Some women choose natural childbirth. They see birth as a natural process that doesn't need painkillers. Relaxation and controlled breathing helps. The father can offer support and encouragement during the birth.

More and more people

There are more people living on Earth today than ever before. The human population is increasing every year. The graph shows how the world population has changed since 1800.

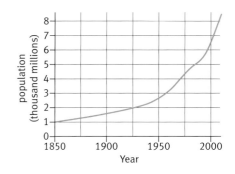

More people can mean more problems

More people need more resources. This puts a lot of demand on planet Earth's resources. In some parts of the world people cannot get enough of the basic resources they need. In other parts of the world people also want more things, like cars, TVs, and clothes. We need to take care of Earth's resources for the next generation.

People make rubbish. This creates pollution.

People need fuel to heat buildings and run cars. This creates pollution, and uses up raw materials.

People need homes and food. Land is cleared for growing crops, and for materials to make houses.

People need clean water, and people make sewage. Cleaning water takes energy.

More people need more resources.

Questions

1 Write a sentence about each of these words:

   sperm   egg   embryo

   fetus   cord   placenta

2 What do you think about these ideas?

   Identical twins are exactly the same.

   People should be more careful with resources for the next generations.

An ammonite fossil.

This fossil was dug out of rock. It is an ammonite. Ammonites do not live on Earth anymore. They are extinct. We can only learn about extinct species from fossils. Fossils are found in ice, amber, and tar pits as well as in rocks.

How are fossils made?

When an animal or plant dies most of its body disappears in days. Microbes rot the soft parts of the body. But if the animal dies near water it may become a fossil. Fossils are found in some types of rocks. These rocks are made of layers of mud. The diagram explains how this happens.

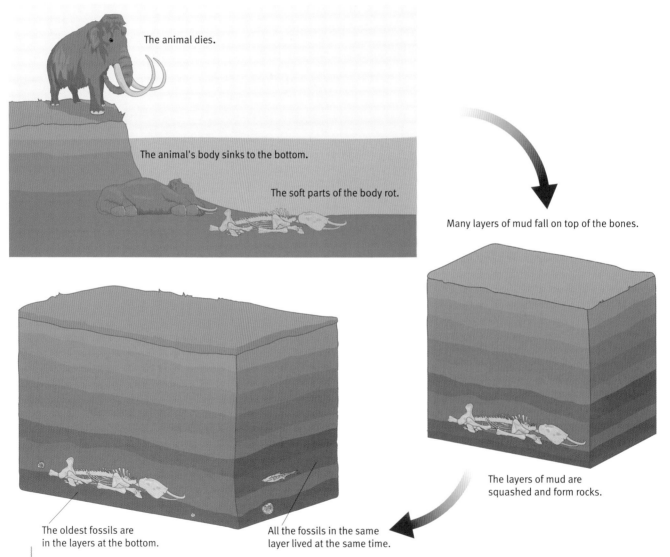

The animal dies.

The animal's body sinks to the bottom.

The soft parts of the body rot.

Many layers of mud fall on top of the bones.

The layers of mud are squashed and form rocks.

The oldest fossils are in the layers at the bottom.

All the fossils in the same layer lived at the same time.

Scientists can work out the age of the rock around a fossil. This tells them how old the fossil is.

Life on Earth

The first living things on Earth consisted of just one cell. They lived about 3500 million years ago.

Over millions of years these simple living things changed so much that they split into different **species**. Living things can only reproduce with other members of the same species. Most of these species did not survive. A few of them kept changing and became the animals and plants on Earth today. This very slow change in living things is called **evolution**.

A few species hardly change in millions of years. Crocodiles are almost the same as when the dinosaurs became extinct. Most species change a lot. Today's birds evolved from dinosaurs. Birds and dinosaurs are very different to each other.

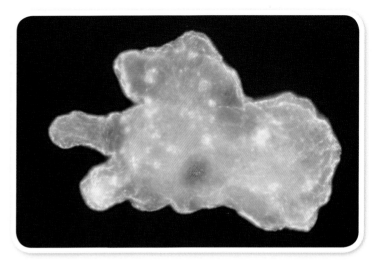

Spot the difference

It's easy to spot differences between dogs and people. But not all dogs are the same. Neither are all people. The differences are called **variation.**

Competition

These chicks are competing for food and water. Chicks that survive will compete for shelter to build their own nest and a mate to breed with.

Crocodiles have hardly changed in millions of years.

Survivors

Sometimes being a bit different can help an animal or plant to survive. Survivors will pass their features on to the next generation. This is how the species slowly changes.

① *Living things in a species are not identical. They have variation.*

Ancestors of modern giraffes had different length necks.

② *Animals compete for things like food, shelter, and a mate. But what if something in the environment changes?*

Food supply became scarce. The giraffes competed for food.

③ *Some will have features that help them to survive. They are more likely to breed. They pass their genes on to their offspring.*

Taller giraffes were more likely to survive and breed. They passed on their features.

④ *More of the next generation have the useful feature. If the environment stays the same, even more of the following generation will have the useful feature.*

Over many generations, more giraffes with longer necks were born.

What causes extinction?

Dinosaurs became extinct about 65 million years ago. Scientists think that the climate on Earth changed suddenly. This changed the places where the dinosaurs lived (their habitats) so much that they could not survive.

So plants and animals may become **extinct** if their habitat changes and they cannot survive.

Pandas live in parts of China where bamboo grows. Each year there is less bamboo in their habitat. Pandas cannot eat anything else.

Dodos were slow-moving. They were easy prey for humans and other animals.

The sabre-toothed tigers' habitat changed. They were also hunted by humans.

Pandas only eat bamboo.

Species may also become extinct if:

- another species in the habitat is better suited to survive there.

Red squirrels used to live all over the UK. Now the larger grey squirrel has taken over most of their habitat.

At risk

Each year there are fewer pandas left on Earth. Pandas are **endangered**. This means that they could soon be extinct. There are lots of other endangered species.

Human beings have caused many species to become endangered or extinct.

Wild primroses in the United Kingdom are endangered. More land is being used for housing and farming. People also take primroses from the wild for their gardens. Taking plants from the wild is illegal.

Gorillas live in Africa. They are endangered because the forests they live in are being cut down. Scientists think that only 10% of their habitat will be left by 2030.

Gorillas are also endangered by hunting. Poachers hunt gorillas illegally.

Year	Population of Northern White Rhinos
1970	500
1980	15
1990	30
2000	10
2010	4

The population of Northern White Rhinos has been dropping for the past 40 years. There are now so few left that they may soon be extinct.

Red squirrels only live in a few places in the UK.

Questions

1 Write a sentence about each of these words:

   species   evolution
   variation   extinct
   endangered

2 What do you think about these ideas?

   Fossils tell the story of life on Earth.

   Human beings should not endanger any other living things.

These paramedics are treating a patient.

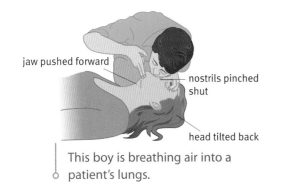

jaw pushed forward

nostrils pinched shut

head tilted back

This boy is breathing air into a patient's lungs.

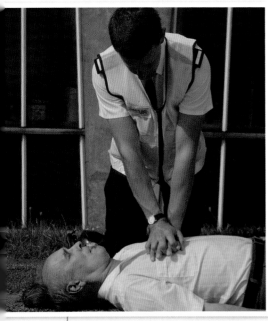

This paramedic is doing the job of the patient's heart.

ABC check

The person next to you suddenly collapses on the floor. He isn't moving. What should you do?

The ABC code tells you what to check first.

A is for airways. Look in his mouth and throat. Is the way clear for air to get in?

B is for breathing. Put your hand in front of his mouth. Is air going in and out?

C is for circulation. Put your head on his chest. Can you hear a heartbeat?

If all three answers are yes, then he isn't in immediate danger. Call an ambulance!

If an answer is no to any of the questions, then you need to help him right away. While you help the casualty get someone else to call for an ambulance.

Kiss of life

Lungs take in air so that **oxygen** can get into the body. If the patient isn't breathing, you need to do it for them. The boy in the diagram checked that the patient's airway was clear. Now he is breathing air into her lungs. He'll keep doing this until she starts breathing.

Heart start

Blood carries oxygen from the man's lungs to the rest of his body. If his **heart** isn't beating, blood stops moving. If oxygen doesn't get to his brain he will die in just a few minutes. The paramedic in this photo is pushing on the patient's chest. This forces blood out of the patient's heart and around his body.

Calling for help

You need to be trained to do the kiss of life, or keep someone's heart pumping. You can learn how at first-aid classes. Knowing this first aid may one day help you save someone's life.

Our patient's life is saved. He is breathing, and his heart is beating. His blood is picking up oxygen from the lungs and carrying it around his body. Now is the time to phone for an ambulance.

- Dial 999 or 112.
- Ask for an ambulance.
- Tell them where the patient is.
- Say what has happened to the patient.
- Give them your name and phone number.
- Stay with the patient until the ambulance arrives.

Rice

Athletes often use the RICE procedure to help with painful muscle injuries and sprains.

- **R**est– don't use the part that hurts.
- **I**ce – cool the injury down with an icepack.
- **C**omfortably support – wrap a bandage tightly around the injury.
- **E**levate – raise the injury up so that blood can flow from it easily.

Keeping the heart beating

You should look after your heart. It is a matter of life and death!

Your heart is a hollow muscle. The heart muscle pumps about 70 times a minute all of your life. All muscles need a good supply of oxygen and food. Small arteries supply the heart muscle with blood. They are called coronary arteries.

If some of these arteries become blocked, part of the heart does not get enough blood. The muscle cells are starved of oxygen and food. Some of the heart cells can die. This is a heart attack. It can stop your heart beating and kill you.

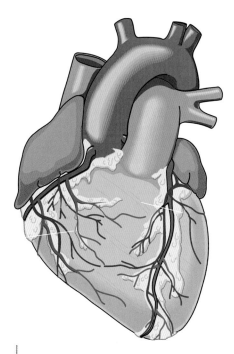

The heart is about the size of a clenched fist.

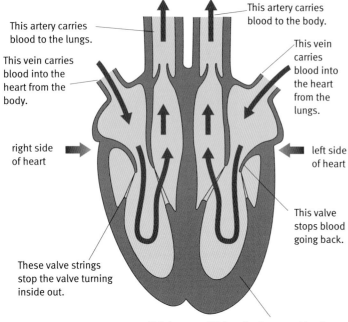

This artery carries blood to the lungs.

This artery carries blood to the body.

This vein carries blood into the heart from the body.

This vein carries blood into the heart from the lungs.

right side of heart

left side of heart

This valve stops blood going back.

These valve strings stop the valve turning inside out.

This has very thick walls; it pumps blood to the body.

Inside your heart

Your heart is made of muscle. It is split into two pumps. The right side of your heart pumps blood to your lungs. The blood picks up oxygen and goes back to the heart. The left side of your heart pumps the blood to the rest of your body.

Blood highway

Blood passes around your body in tubes called blood vessels.

- **Arteries** carry blood away from your heart. Blood comes out of the heart under a lot of pressure. Arteries need thick walls so that this high pressure does not damage them.
- **Veins** carry blood towards your heart. The blood pressure is low, so veins can have thin walls. Blood moves slowly in the veins.
- **Capillaries** are tiny tubes connecting the arteries to the veins. They are everywhere in your body. Capillaries have very thin walls so food and oxygen can easily pass from the blood into your cells.

Blood loss

The blood vessels close to your skin are capillaries. These carry blood at low pressure so you won't lose much blood from a small cut. A deep cut may get into an artery. This is much more serious. High-pressure blood will quickly escape from the cut unless it is closed again.

You can lose up to 10% of your blood without any harm. A loss of 30% is serious. You will probably need a transfusion of fresh blood if you lose this much.

Valves in your heart work like one-way gates. They make sure that blood does not flow backwards in the heart.

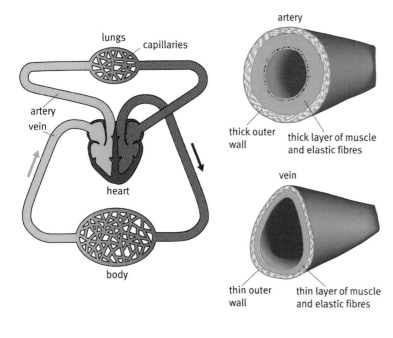

lungs

capillaries

artery

artery

vein

heart

body

thick outer wall

thick layer of muscle and elastic fibres

vein

thin outer wall

thin layer of muscle and elastic fibres

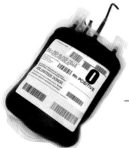

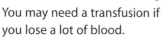

You may need a transfusion if you lose a lot of blood.

Healthy hearts

In the UK about 270 000 people have a heart attack each year. There are things you can do to cut your risk of a heart attack.

Risky business

Paul knows that doctors say he shouldn't smoke. But he's a bit confused. Paul says, 'My grandfather has smoked since he was 16. He's 78 now, and he isn't ill.'

Smoking increases your risk of lots of illnesses – heart attacks, cancers, breathing illnesses. This doesn't mean that everyone who smokes will have a heart attack or get lung cancer. But if you smoke you are more likely to get these illnesses. Stopping smoking cuts your risk of getting ill straight away.

Paul wonders about the risks of smoking.

Cigarettes smoked per day	Number of cases of cancer per 100 000 men
0 – 5	15
6 – 10	40
11 – 15	65
16 – 20	145
21 – 25	160
26 – 30	300
31 – 35	360
36 – 40	415

Questions

1 Write a sentence about each of these words:

   lung   **oxygen**   **heart**   **artery**
         vein   **capillary**

2 What do you think about these ideas?

   Everyone should be made to learn first aid.

   If someone has a heart attack, it is their own fault.

What am I made of?

You are made of six different types of chemical. Each chemical has its own job in your body. You get these chemicals from your food and drink. Most of your body is water.

Almost everything you eat is a mixture of these different food chemicals.

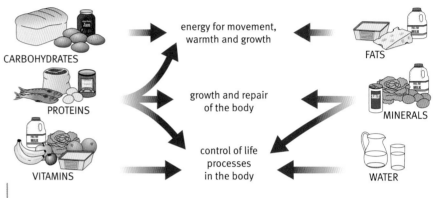

The nutrients in food and what they do.

Chemical	Job
carbohydrate	provides energy
protein	builds cells
fat	stores energy
fibre	provides bulk
minerals	build teeth and bones
vitamins	keep you healthy
water	dissolves other chemicals

Where does food come from?

Your food comes from plants and animals.

Plants supply you with **carbohydrates** for energy:
- plant seeds (wheat, barley, maize, rice)
- vegetables (potatoes, carrots, yams)
- fruit (bananas, oranges, apples, grapes)
- sugars (cane, beet, milk, honey)

You can get **protein** for growing and repairing your body from plants and animals:
- meat (chicken, beef, lamb, salmon)
- milk (cheese, yogurt)
- eggs
- plant seeds (peas, beans, lentils)

You need fat in your diet to give you a store of energy:
- meat (beef, lamb, pork, chicken)
- nuts (walnuts, peanuts)
- plant seeds (olive, almond, sunflower)
- milk (butter, cream, yogurt, cheese)

You need fibre in your diet to give your food bulk. This helps your muscles push the food along:

- baked beans
- whole-wheat cereals

Typical value per 100g	
ENERGY	1550 kj
PROTEIN	7 g
CARBOHYDRATES	83 g
of which sugars	7 g
starch	76 g
FAT	1 g
of which saturates	1 g
FIBRE	0.2 g
SODIUM	2.5 g
VITAMINS	20 mg
IRON	7.9 mg

Food labels

Take look at this food label. It is from a packet of breakfast cereal.

The label tells you that there is lots of carbohydrate in the cereal. It also tells you that there is not much protein. So cereal is an energy food, not a body-building food. The label tells you how much energy you will get from a bowl of cereal.

Diet

Not everybody needs to eat the same food. It all depends on how you live.

I need a lot of fat in my diet to keep me warm.

I eat plenty of protein, minerals, and carbohydrates for my baby.

I always eat lots of carbohydrate for energy before a race.

My mum says I need meat to give me protein. This will help me grow up strong.

Deficient diets

Some people on Earth have a diet with enough carbohydrate, protein, **minerals**, and **vitamins** for an active, healthy life.

Many people are starving. They don't get enough energy in their food, so they may not be very active. They don't have enough protein in their diet. This means that they may not grow properly. Shortage of vitamins may cause other illnesses.

Other people choose to eat a poor diet. They may leave out fruit, which may increase their risk of cancer. Or they may eat food with more energy in it than they use up in exercise. The extra food is stored as fat. Being overweight increases a person's risk of heart disease.

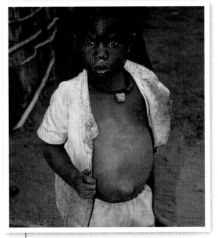

This child has an illness called Kwashikor. He does not get enough protein in his diet.

A good diet isn't the only important thing for a healthy life. Regular exercise is also important. It keeps all of your body systems in top condition.

Digestion

Your food contains many different chemicals. Most of these chemicals have large molecules. Large molecules are no use to your body because they cannot dissolve. We say that they are insoluble. Insoluble molecules cannot pass from your gut into your blood.

Large food molecules must be broken down so that they can get into your blood. Then they can be carried to your cells that need them. Food is broken down in your digestive system.

Breaking large food molecules into smaller ones is called digestion. It has three stages:

- When you chew your food you break it into smaller pieces that make it easier to digest.
- The food mixes with special chemicals called enzymes. These speed up the breakdown of large molecules into smaller ones.
- The small molecules pass through the wall of the small intestine. They are absorbed into the blood.

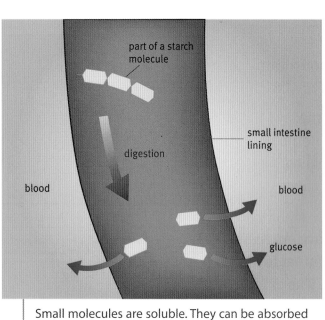

Small molecules are soluble. They can be absorbed into your blood.

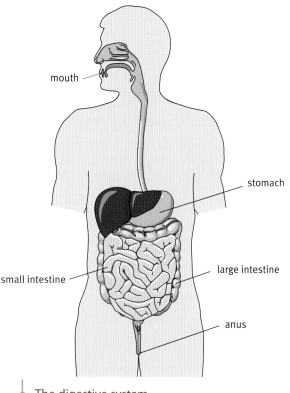

The digestive system.

Enzymes

Enzymes are a special group of chemicals in your body. They speed up chemical reactions in your cells. Many of the reactions that keep you alive will only happen if the correct enzyme is present.

Some enzymes help with digestion. Enzymes act like chemical scissors. They cut the long food molecules into shorter ones, over and over again. Each type of food molecule needs its own special enzyme. They are found in different parts of the digestive system.

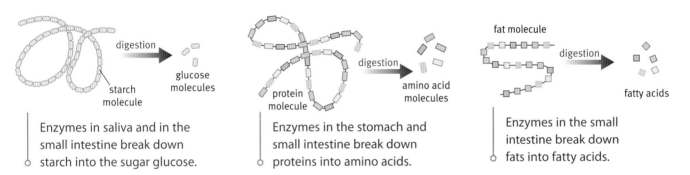

Enzymes in saliva and in the small intestine break down starch into the sugar glucose.

Enzymes in the stomach and small intestine break down proteins into amino acids.

Enzymes in the small intestine break down fats into fatty acids.

Enzyme	Food digested	Digestion product	Location
amylase	starch	glucose	starts in the mouth, finishes in the small intestine
protease	protein	amino acids	starts in the stomach, finishes in the small intestine
lipase	fat	fatty acids	small intestine

Questions

1 Write a sentence about each of these words:

carbohydrate   protein   fat

mineral   vitamin   enzyme

2 What do you think of these ideas?

Everything you buy to eat should have a food label on it.

Eating a variety of foods in moderation helps you live longer.

If your core temperature drops below 35 °C you can become unconscious and die.

Shivering warms you up.

If your core body temperature gets too high or too low it makes you very ill. Your core body temperature must be kept steady. Other things inside your body must also be kept steady, like the amount of water and sugar in your blood. Your body has control systems to do this.

Body temperature

Your body works best at 37 °C. Your brain is always checking the temperature of your blood. If it goes above or below 37 °C, your brain sends out signals to the rest of your body. Your skin and muscles take action to get your temperature back to 37 °C.

Replacing heat

Your body temperature is 37 °C. Heat always flows from warmer places to cooler places. The room you are in is probably about 20 °C, so heat is flowing steadily out of you into the room. To stay at 37 °C your body must replace this heat.

When your muscles work they give out heat. When you are moving, heat from your muscles warms you up.

When you are cold you **shiver**. Muscles under the skin contract. This gives out extra heat to help you warm up. Exercise such as running around or beating your arms has the same effect.

Keeping warm

When you go outside in winter you usually put on extra clothes. Clothes stop you from losing too much heat.

Skin

Your skin also has ways to stop your body losing too much heat:

- Layers of fatty tissue provide heat insulation.
- Raising the hairs stops air carrying heat away.

Clothes slow down the loss of energy from your skin.

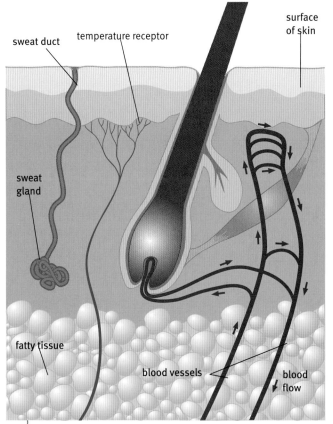

Cross-section through the skin.

Cooling down

When you are hot you **sweat**. Water evaporates from your skin and takes heat away from you.

Small blood vessels carry blood just under the surface of your skin. When you are hot more blood flows through these blood vessels. This makes it easier for heat to pass from your blood into the air. This is why your face looks redder when you are hot.

Water

Water is going in and out of your body all the time. It goes in as food and drink. Doctors recommend that you drink two litres of water every day to stay healthy. You also make a lot of water as a waste product of respiration.

Inputs
water content in:
• food
• drink
water made in:
• respiration

Outputs
water content in:
• exhaled air
• sweat
• urine
• faeces

You lose water in several ways. Some water leaves with the air you breathe out. Some water leaves your skin as sweat. The rest leaves as urine and in your faeces.

Kidneys

You have two **kidneys**. They filter your blood to take out waste chemicals such as urea. They also filter water out of your blood. Water and urea make urine. The urine is stored in your **bladder** until you excrete it.

Water control

Keeping the right amount of water in your body is critical. Your brain senses the dilution of your blood – how much water there is. When your blood is too dilute, your kidneys remove water from it.

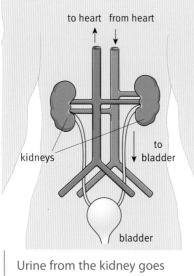

to heart   from heart

to bladder

kidneys

bladder

Urine from the kidney goes to the bladder.

Controlling blood sugar

You eat a large meal. As the food passes through your gut, the carbohydrate is broken down into sugar (glucose) and enters your blood.

The level of sugar in your blood goes up. If it gets too high your pancreas makes more of a chemical called **insulin**. Insulin is a hormone. It is carried around your body in your blood.

Insulin makes your muscles take up glucose to use it for energy. Insulin also lets the liver take sugar out of the blood and store it. Your blood sugar level goes back down to normal.

Diabetes

Jo has had diabetes since she was 12 years old. Her pancreas can't make insulin any more. Without treatment her blood sugar level would be much too high. This is very dangerous.

Jo can control her diabetes. Before she eats a meal she injects herself with insulin. She is also careful not to eat too much sugary food.

Sometimes after a meal Jo can find herself getting bad-tempered. Her blood sugar is too low because she had too much insulin. Once or twice she has nearly passed out. 'I always have some peppermints in my pocket,' says Jo. 'I can eat one if I feel my glucose levels are too low.'

In type-1 diabetes the special cells in the pancreas are destroyed. The pancreas cannot make insulin.

In type-2 diabetes the pancreas does not make enough insulin or cells do not respond to the insulin there is.

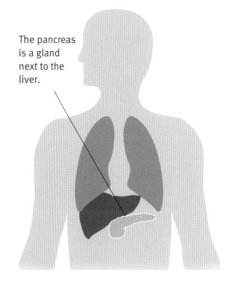

The pancreas is a gland next to the liver.

Insulin controls the level of sugar in blood. It lets sugar molecules into cells.

When the sugar levels rise the pancreas cells release insulin into the blood.

A person with diabetes injects insulin several times a day. This keeps the blood sugar level normal.

Questions

1 Write a sentence about each of these words:

   shiver   sweat   kidney   bladder   insulin

2 What do you think about these ideas?

   Old people shouldn't have to pay their winter heating bills.

   You should always drink water when you feel thirsty.

Lungs

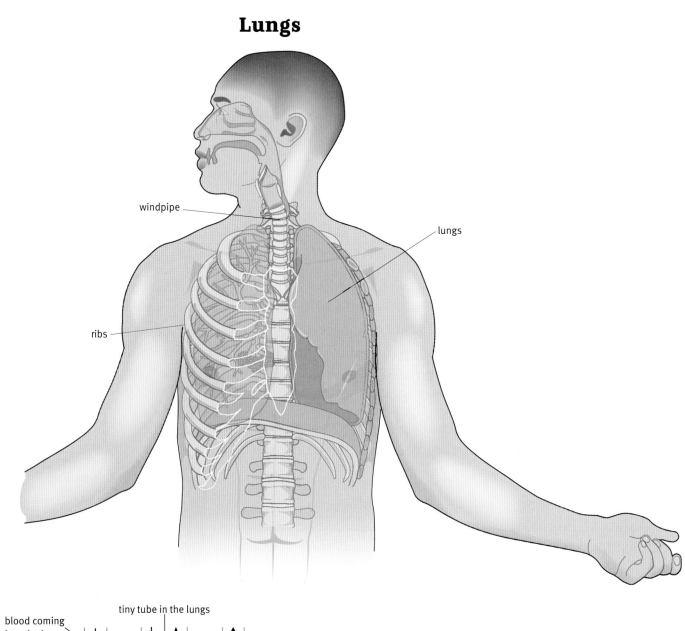

windpipe

lungs

ribs

blood coming into the lungs

tiny tube in the lungs

blood leaving the lungs

oxygen passes into the blood

carbon dioxide passes into the air

air sac

If you flattened out all the air sacs in your lungs they would cover a football pitch.

When you breathe in, air goes into your **lungs.** Tubes carry air deep into the lungs. The tubes divide into smaller and smaller tubes.

At the end of each tube is a tiny bag (or 'sac') of air. These air sacs are covered in tiny blood vessels.

Here two things happen:
- Oxygen passes from the air into your blood.
- Carbon dioxide passes from your blood into the air.

Breathing in and out

You use muscles in your chest to breathe in and out.

To breathe in:
- the muscles between your ribs pull the ribs up and out
- the diaphragm pulls down.

This makes more space in your chest. Air pushes into your lungs.

To breathe out:
- the muscles between your ribs relax so the ribs move down and in
- the diaphragm relaxes and moves up.

This makes less space in the chest. Air is pushed out of your lungs.

People have different sized lungs. The size is called their lung volume. Smoking can reduce your lung volume. You cannot get as much air into your lungs.

Smokers may also take longer to breathe out than non-smokers. This means it takes them longer to get fresh air into their lungs each time they breathe.

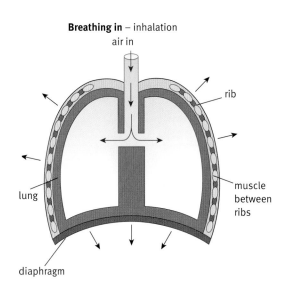

Breathing in – inhalation
air in

rib

muscle between ribs

lung

diaphragm

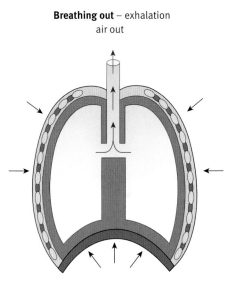

Breathing out – exhalation
air out

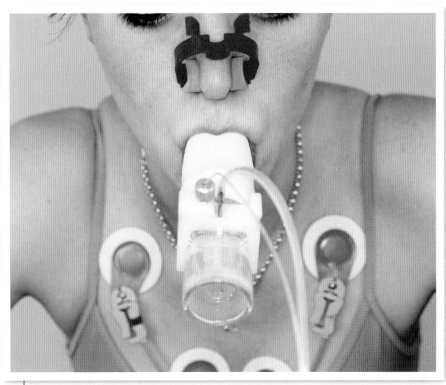

This athlete is having her lung volume measured.

Car exhaust fumes can pollute city centres.

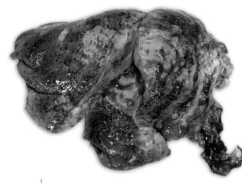
Lung tissue blackened by tar from cigarette smoke.

Health warning in 2003.

You can get ill by breathing in smoke from other people's cigarettes. This is called passive smoking.

Asthma attack!

Asthma also affects breathing.

What triggers asthma?

Car exhaust fumes pollute the air. In December 1991 London had very unusual weather. There was no wind to blow away the exhaust fumes. So the level of air pollution was very high.

The number of people dying each day from asthma rose by 50%. When the weather changed the air pollution level fell. The number of deaths from asthma dropped to normal.

Maybe the car exhaust fumes caused the asthma deaths. Maybe it was just a coincidence. It is not easy to be sure. What we do know is that car exhaust fumes contain pollutants that can make asthma worse.

Smoking

Cigarettes can kill. They can take years off your life. Smoking 20 cigarettes a day multiplies your chances of dying of lung cancer by 15 times. Smoking also causes cancer of the throat, bladder, and mouth.

The cancer is caused by tar. This gets into your body through the lungs. Tar reduces your lung volume and peak flow rate.

Tobacco smoke also contains carbon monoxide. This is a poisonous gas. It reduces the amount of oxygen your blood can carry. It is called the silent killer because you can't see it or smell it.

Nicotine in tobacco smoke is addictive. It gives you a buzz. But it also increases your risk of heart disease.

Some people use patches to help them give up their cigarette habit. Nicotine in the patch slowly seeps through the skin into the blood. This gives them the buzz without the smoke.

Using oxygen

Brain cells, skin cells, fat cells ... the cells in your body can only do their job if they have oxygen.

Air you breathe in has oxygen in it. Blood carries the **oxygen** to your cells.In your cells oxygen reacts with **glucose**. This is called **respiration**.

Respiration releases energy from the glucose. Your cells use this energy to do their job.

Respiration makes the waste products **carbon dioxide** and water. Blood carries the carbon dioxide to your lungs.

You breathe out the carbon dioxide. Some of the water is also breathed out.

Working harder

This athlete has worked his muscles very hard. The muscles need extra energy to do this work. The muscle cells need more oxygen and glucose. They are making more carbon dioxide than normal. This carbon dioxide needs to be be got rid of.

Your brain is always checking the level of carbon dioxide in your blood. When you exercise, your brain senses that there is more carbon dioxide in your blood. It sends signals to the heart and lungs to make them work faster.

Blood moves faster through your lungs, and the unwanted carbon dioxide is breathed out faster. You only stop panting when the level of carbon dioxide in your blood is back to normal.

Questions

1 Write a sentence about each of these words:

lung   asthma   oxygen

glucose   respiration   carbon dioxide

2 What do you think of these ideas?

Smoking cigarettes always results in an early death. Asthma is caused by car exhaust fumes.

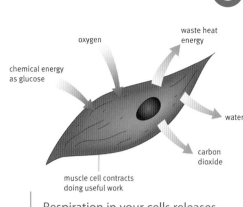

Respiration in your cells releases energy from glucose and oxygen.

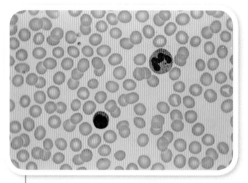

Red blood cells in your blood carry oxygen everywhere in your body.

This athlete is out of breath!

These sunflower plants are making food in their leaves.

Plant food

Plants make their own food. They use carbon dioxide, water, and sunlight. This is called photosynthesis. Photosynthesis takes place in leaves.

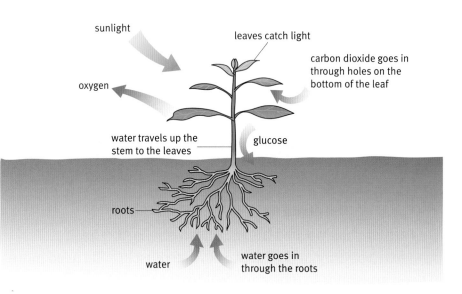

This equation shows what happens in photosynthesis:

$$\text{water} + \text{carbon dioxide} \xrightarrow{\text{sunlight}} \text{glucose} + \text{oxygen}$$

Animal food

Oxygen is a waste product of photosynthesis. So all life on Earth depends on plants for their oxygen. Animals also depend on plants for their food.

Some animals get their food straight from plants. Some animals eat other animals.

Food chain

Leaves are part of plants. They make their own food.

The caterpillar gets its food from eating leaves. This makes it a **herbivore**.

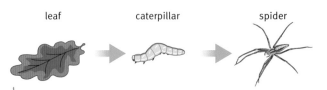

The arrows on a food chain show what eats what.

Rain forests are sometimes called the "lungs of Earth". They make oxygen in photosynthesis.

Spiders eat caterpillars. This makes the spider a **carnivore**.

Spiders, caterpillars, and leaves make a food chain.

Food is passed from the leaf to the spider, through the caterpillar.

Food web

Most animals eat more than one thing. So they are part of more than one food chain.

Food chains join up in a food web. This food web shows what eats what in our countryside.

The bottom layer of the food web has the plants.

The next layer has the herbivores.

The top layers have the carnivores.

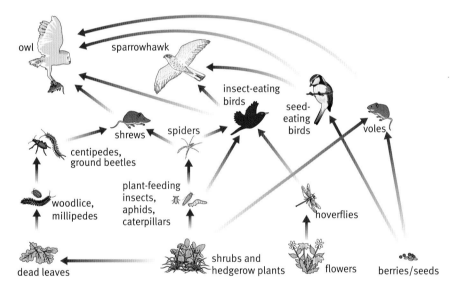

Changes in a food web

The photo shows a mink. Mink eat voles.

Mink escaped into our countryside in the 1980s.

The number of voles went down.

Some other changes also happened in the food web:
- There were fewer voles feeding on the shrubs.
- This left more food for the caterpillars.
- Numbers of caterpillars went up.

Mink escaped from farms. They were bred for their fur to make mink coats.

Predator or prey?

Lions hunt antelope. The lion is a **predator**. The antelope is its **prey**.

Most of the time, the antelope gets away.

This is because the lion is running for its lunch, but the antelope is running for its life!

Adapt and survive

An antelope has features that help it survive being caught. These features are called adaptations. Here are some of the adaptations:

- it can fight with its horns
- it can run very fast
- it has all-round vision to watch out for predators.

Here are some adaptations that make the lion a good predator:

- it has sharp teeth and claws
- it has strong jaw muscles
- it has forward-facing eyes for judging distance.

Looking at habitats

Where an animal or plant lives is called its **habitat**. You can investigate what lives in a habitat with the equipment below.

How many?

Ali grows wheat. He uses a quadrat to count weeds in part of his field.

'I throw the quadrat in different parts of the field,' says Ali. 'Then I count the different weeds in the quadrat.'

Ali could use a simple key to identify the weeds in his field.

In this habitat weeds compete with the crop for space, light, and water.

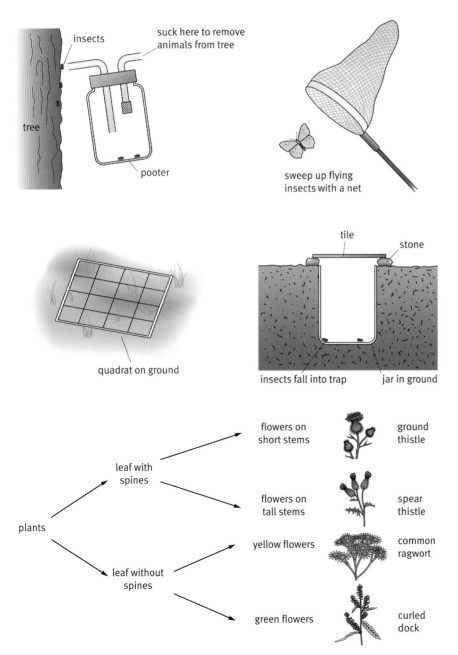

insects

tree

suck here to remove animals from tree

pooter

sweep up flying insects with a net

quadrat on ground

tile

stone

insects fall into trap

jar in ground

plants

leaf with spines

flowers on short stems — ground thistle

flowers on tall stems — spear thistle

leaf without spines

yellow flowers — common ragwort

green flowers — curled dock

Questions

1 Write a sentence about each of these words:

herbivore   carnivore   predator   prey   habitat

2 What do you think about these ideas?

We need to know about all the plants and animals that live where we live.

It is more natural to be a vegetarian than a meat eater.

One day a fast response could save your life.

Your **senses** tell you about your surroundings. You are sensing changes around you all the time. Sights, sounds, hot, or cold – all these changes are called stimuli. Your body reacts to these changes. The reaction is called your response.

Some responses happen automatically. You don't have to think about them. Automatic responses are called simple **reflexes**. Simple reflexes protect you from getting hurt.

How do reflexes work?

You touch a very hot plate. You quickly move your hand away. This simple reflex protects you from damage.

Special cells in your skin detect the hot plate – the stimulus. These cells are called sensors or **receptors.** Muscles in your arm cause the response. Muscles are the **effectors.**

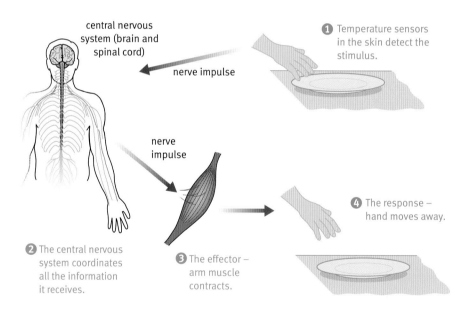

central nervous system (brain and spinal cord)

nerve impulse

1 Temperature sensors in the skin detect the stimulus.

nerve impulse

2 The central nervous system coordinates all the information it receives.

3 The effector – arm muscle contracts.

4 The response – hand moves away.

Nerve cells link different parts of the body together (magnified about 600 times).

axon

Skin sensors

Your skin has different types of sensors. Each sensor is at the end of a nerve that carries signals towards your brain. Sensors for touch, pain, heat, and cold are close to the surface of your skin. Pressure sensors are much deeper. Some areas of your skin have more sensors than others. Your fingertips have the most sensors.

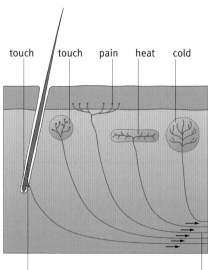

touch   touch   pain   heat   cold

hair

nerve fibres to spinal cord and brain

Pain sensors are nearer the surface of the skin than pressure sensors.

The pupil reflex

Imagine you walk from a dark cinema into a light room. The bright light could damage your eyes. A simple reflex stops this from happening:

- Sensor cells in your eyes detect the light.
- Nerves take signals to your brain.
- Nerve cells take signals to muscles in your eyes.
- The muscles are the effectors.
- The muscles make your pupils smaller.
- Less light gets into your eyes.

The human eye

Your eyes give you information about light around you. Each part of the eye does a different job:

- The cornea is clear so light passes through.
- Some light goes through the pupil, a hole in the middle of the iris.
- The lens guides the light onto the back of the eye.
- Sensor cells in the retina detect the light.
- Signals from the retina pass along the optic nerve to your brain.

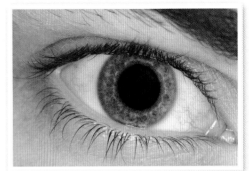

These photos show a simple reflex. The stimulus is bright light. The response is the pupils getting smaller.

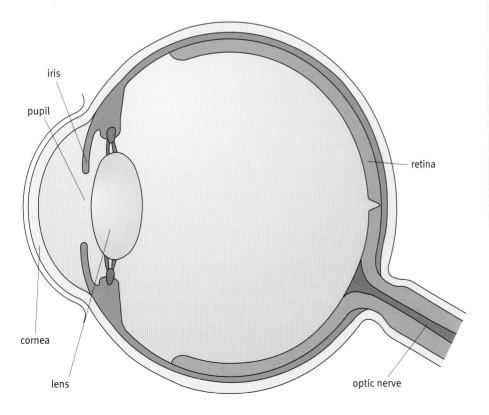

iris

pupil

retina

cornea

lens

optic nerve

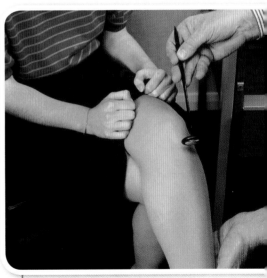

The knee-jerk reflex makes your thigh muscle contract, so your leg straightens. Doctors may test this reflex when you have a health check. Try standing still with your eyes closed. You will notice this reflex helping you to balance.

Seeing in three dimensions

A carnivore has binocular vision.

Herbivores have monocular vision.

Birds have monocular vision, except for birds of prey like this owl.

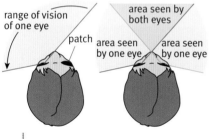

range of vision of one eye
area seen by both eyes
patch
area seen by one eye
area seen by one eye

Binocular vision.

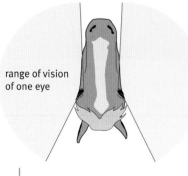

range of vision of one eye

A horse has a wide field of view.

You have two eyes in your head. This is not in case one of them stops working!

You can see up, down, and sideways. Two eyes make it possible for you to see the world in three dimensions.

Binocular vision

Your eyes are both at the front of your head. This gives you binocular vision. Part of what your eyes see overlaps. This helps you to judge how far away something is.

Binocular vision is useful for predators. They can work out the distance between them and their prey.

Monocular vision

Many prey animals have an eye on each side the head. This is called monocular vision. Prey can't judge distances so easily, but they have a much wider field of view. They can see more of the space around them. This helps them to spot the predators coming.

Taste, smell, flavour

Your eyes tell you a lot about your food. But looks can lie. Sensors on your tongue and in your nose are much more reliable.

Your tongue is covered in taste buds. These contain sensor cells that detect tastes. You have four types of sensor for taste. They tell you if your food is salty, sour, sweet, or bitter. Each time a chemical hits the right sensor a signal passes along a nerve to your brain. Then your brain knows something about the food in your mouth.

Your nose is lined with about 200 different types of smell sensors. Each type picks up different chemicals in the air. They give you much more information about the flavour of your food, and the smells around you.

You can't smell your food when a cold blocks your nose. The food loses a lot of its flavour.

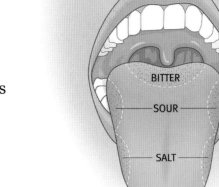

Your tongue has taste sensors on its surface.

Dogs have a very sensitive sense of smell.

Questions

1 Write a sentence about each of these words:

sense   reflex   receptor

effector   nerve

2 What do you think about these ideas?

We would be better off with eyes at the side of our head.

You can't taste food very well when you have a cold.

Plant food

Plants use air, water, and sunlight to make sugar in their leaves. This is their food. Most of it is stored as **starch** in the leaves, roots, and seeds. When you eat plants, this starch becomes your food.

Wheat is seeds.

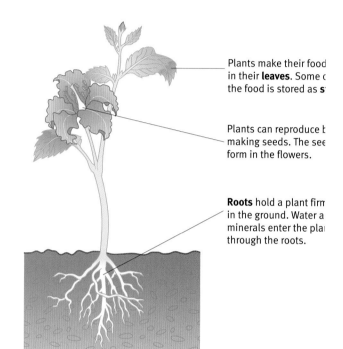

Plants make their food in their **leaves**. Some of the food is stored as **s**

Plants can reproduce by making seeds. The seeds form in the flowers.

Roots hold a plant firmly in the ground. Water and minerals enter the plant through the roots.

Starting from seed

New plants grow from **seeds**. When the conditions are just right the seed begins to grow. This is called germination. Seeds need warmth, air, and water to grow.

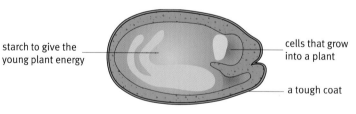

starch to give the young plant energy

cells that grow into a plant

a tough coat

Growing cuttings

Emma grows plants to sell to garden centres.
- She uses a plant she likes and takes a cutting.
- She dips the cutting into rooting powder.
- She puts the cutting in compost.
- The cutting soon grows new roots to make a new plant.

All of the new plants come from the same parent. They all have exactly the same genes, so they are identical. They are **clones**.

Tubers and runners

Many plants can make their own clones. This allows them to spread and survive from year to year.

- Some plants make tubers under the ground. These contain starch, which provides food for the new plant while it is growing its leaves.
- Some plants send out runners across the ground. Their ends can take root and make new plants.

The cloned plants may not look exactly the same if they are grown in different conditions. They may get different amounts of water or light. They may get different amounts of nutrients from the soil.

Soil

Plants grow in soil. But soil is not all the same. Some soils are better than others. They are more fertile. Farmers can grow more food in a fertile soil. They get a better yield.

A good soil should have:
- water, but not too much
- plenty of air between the clumps of soil
- plenty of nutrients, which plants need to grow.

Farmers can look after their soil by:
- ploughing their fields to break up the soil
- clearing ditches to stop the soil becoming waterlogged
- adding plant material (humus) to put in nutrients and stop the soil drying out
- adding manure or **fertiliser** to put in nutrients.

Potato plants make lots of tubers full of starch under the ground. In the spring, fresh plants grow from each potato.

This spider plant has put out several runners. Each runner can become a separate plant when it roots in soil.

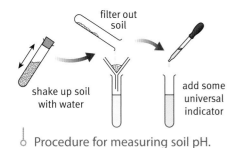

Procedure for measuring soil pH.

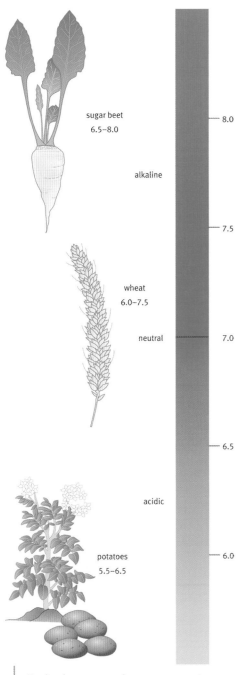

Each plant grows best at a certain soil pH.

sugar beet
6.5–8.0

alkaline

wheat
6.0–7.5

neutral

acidic

potatoes
5.5–6.5

8.0

7.5

7.0

6.5

6.0

Acid or alkali?

Some soils are acidic. Some are alkaline. You can measure the pH of soil. Each plant grows best in soil that has a certain pH. To get the best yield farmers must grow plants in soil with the right pH.

Growing healthy plants

A healthy plant needs a good supply of certain chemicals from the soil. These nutrients are dissolved in water in the soil. They get into the plant through the roots.

Many farmers use fertilisers to add nutrients to the soil. This increases the yield of their crops. A fertiliser usually contains nitrogen, potassium, and phosphorous.

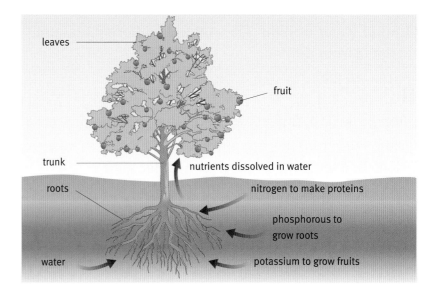

Organic farming

Farmers who grow **organic** food aren't allowed to use chemical fertilisers. They use natural fertiliser (manure) instead. Dung from animals is spread on the land.

Organic farmers do not use pesticides. They rely on crop rotation. They don't grow the same crop in a field year after year. Pests that attack one crop have died out before the same crop returns to the field.

Is organic food better? Some people think it tastes better. It certainly contains far fewer pesticide residues, which some people worry about. Organic farming is certainly better for the soil.

Milk

Most of the milk we drink comes from cows. Milk products can also be made from sheep or goats' milk. Des has a herd of cows. Some of his cows make more milk than others.

"I want all of my cows to make lots of milk," says Des. "I'm only going to keep the calves of the cows that make the most milk."

Des is using **selective breeding**. He hopes the cows with lots of milk will pass this feature on to their calves. Of course he has to choose the right bull as well – one whose mother made a lot of milk is best.

Bacteria control

Milk is a great food for young mammals. It is also great food for some types of bacteria.

Here are three ways to stop milk going off:
- Refrigeration: cool milk to 4 °C. This slows down the bacteria. Milk will stay fresh for up to eight days in a fridge.
- Pasteurisation: heat the milk to 70 °C for only 15 seconds. Then seal it up. This kills only the harmful bacteria. The taste of the milk is the same. It will keep for up to 20 days in a fridge.
- Sterilisation: heat the milk to 135 °C to kill all the bacteria. Then seal it up. It will keep fresh for several months, even out of a fridge, but it won't taste the same.

Milk is often tested for its bacteria content before it goes to the shops. This is to make sure that it is safe to drink.

Questions

1 Write a sentence about each of these words:

   starch   clones   organic   fertiliser   selective breeding

2 What do you think about these ideas?

   You could survive on a diet of milk alone.

   Organic farming is better for the planet.

Production

Milk is produced on the farm.

Transportation

It is transported to the dairy.

Processing

Milk is pasteurised and homogenised.

Storage

Milk can be heat-treated or dried for storage.

Delivery

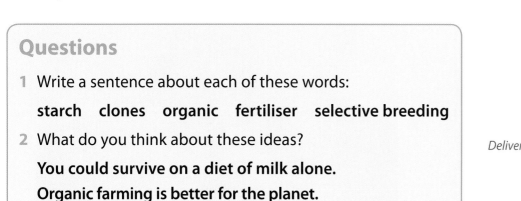

Milk can be delivered fresh to our door or supermarket, or made into butter, cheese, yoghurt, cream, and powdered milk.

Mick and Jane do drugs together.

Drugs everywhere

Mick and Jane often eat out together. This involves taking drugs:

- They have alcohol in the wine they drink with the food.
- They have caffeine in the coffee at the end of the meal.
- Mick inhales nicotine from a cigarette when they leave the restaurant.
- Jane takes paracetamol the next day for a headache.

All four drugs are legal. The paracetamol is beneficial. The nicotine is definitely harmful. The other two may be harmful if they take too much.

A drug is any chemical that affects the mind or body. There are lots of drugs. Some are good for you, others are not. Some of them are illegal.

Taking drugs

There are many different ways you can take drugs:

- Drink alcohol in wine or beer and caffeine in tea or coffee.
- Swallow medicine tablets such as aspirin or paracetamol.
- Inject illegal drugs such as heroin or steroids.
- Absorb nicotine through the skin using patches.
- Inhale solvents or cannabis.

Inhaling a drug.

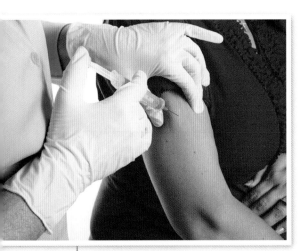
Injecting a drug.

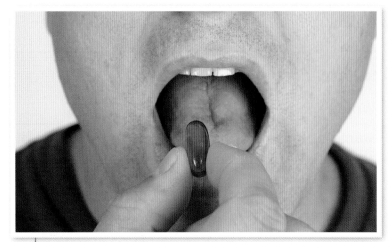
Swallowing a drug.

Addicted

Mum's got lung cancer. It's horrible. Will I get it too?

You might, if you don't stop smoking. The smoke gives you cancer.

I can't give up. I have to have a cigarette every couple of hours.

You're addicted to the nicotine in the smoke. It's changed your brain.

So I'm going to die of lung cancer?

Not if you switch to nicotine patches. They'll give you the nicotine without the smoke.

Heroin is illegal because it is so easy to get addicted to it. Some legal drugs can only be obtained with a prescription from the doctor. This stops people overdosing as well as getting addicted.

Sport

Athletes take drugs. They do this to help them win.

- Caffeine is a legal **stimulant**. It increases activity in the brain, making athletes more alert.
- Paracetamol is a legal **painkiller**. It blocks nerve impulses in the brain, allowing athletes to ignore injuries and carry on.
- Steroids are illegal **performance enhancers**. These help athletes to grow large muscles. It can also damage their liver.

Professional athletes have their blood tested for illegal drugs. They can be banned from competing if they test positive.

Marion Jones winning the 200-m sprint in the Sydney 2000 Olympics. She won five medals altogether, three of them gold. All were taken away from her when she tested positive for steroids.

Recreation

Lots of people take drugs for fun.

- Nicotine and caffeine are legal stimulants. They excite you.
- Ecstasy is an illegal stimulant. It makes you more sociable.
- Alcohol is a legal **depressant**. It slows down brain activity. This relaxes you.
- Cannabis is an illegal **hallucinogen**. It makes you experience the world differently.

These energy drinks contain a lot of caffeine.

Drinking and driving often leads to death.

Two men enjoying the destruction of their livers?

Alcohol abuse

Alcohol kills more people than any other drug. It can kill them in many ways:

- It can damage their liver or give them cancer.
- It can slow down their reactions so that they have a car accident if driving.
- It can impair their judgement so that they are more likely to get into a fight.

Other effects of alcohol are less life-threating but just as unpleasant:

- slurred speech
- poor balance
- blurred vision
- loss of consciousness.

Legal drugs

Many drugs are legal. These include:

- caffeine
- alcohol
- nicotine
- paracetamol
- aspirin.

Just because they are legal doesn't mean they are safe. Some are addictive, others are harmful. Most unsafe drugs can only be obtained with a doctor's prescription. Alcohol is a notable exception.

Most unsafe legal drugs can only be obtained on prescription.

Illegal drugs

Some drugs are illegal. This may be because they are very addictive or because they may be harmful. Most illegal drugs are recreational – people take them for the buzz. There are three classes of illegal drug in the UK:

- class A drugs include heroin, Ecstasy, and LSD
- class B drugs include cannabis and amphetamines
- class C drugs include ketamine and tranquilisers.

Class A drugs are the most harmful. You will probably go to prison if you are found with a Class A drug. You may only get a police caution if you are caught with a Class C drug. This is because they are less harmful than Class A drugs.

When a drug dealer is caught they go to prison. This is because they harm lots of people. But if you only use drugs, you will receive a lighter sentence. This is because you are only harming yourself.

Both of these drugs are legal, harmful, and readily available.

The only legal drug in this photo may be the most harmful.

Questions

1 Write a sentence about each of these words.

stimulant   painkiller   performance enhancer depressant   hallucinogen

2 What do you think about these ideas?

Alcohol should only be available with a doctor's prescription.

All harmul drugs should be legalised and taxed like cigarettes.

James looks a lot like his parents. His body contains genes he has inherited from them. Genes control how your cells develop and what they do.

All human beings have lots of the same genes – we're 99.9% alike. But your mixture of genes is unique – you're a one off.

Where are genes?

Your body is made of billions of cells. Each cell has a **nucleus**. Inside the nucleus are long threads called **chromosomes**. Each chromosome is made up of thousands of **genes**. Each gene is made of a chemical called DNA.

James and his family.

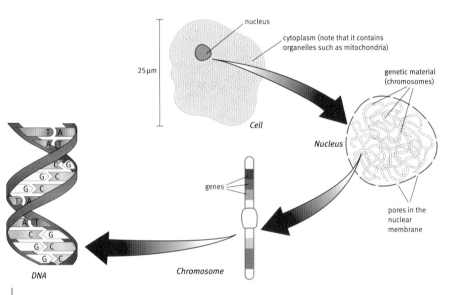

nucleus

cytoplasm (note that it contains organelles such as mitochondria)

25 μm

Cell

genetic material (chromosomes)

Nucleus

genes

pores in the nuclear membrane

DNA

Chromosome

Each **chromosome** in the nucleus contains thousands of **genes**.

How do genes control your cells?

Each length of DNA in your genes contains instructions in a code. Genes are instructions for cells. They tell the cell how to make proteins. Proteins are very important chemicals for cells.

What else controls what I'm like?

Sharma has brown eyes. She has inherited this feature from her parents.

Sharma speaks with an accent. This is not inherited. She got her accent from listening to people talking. So this feature is caused by her environment.

Sharma is a good runner.

Sharma is a very good runner. This feature is partly inherited. But it is also partly caused by her environment. Sharma has trained very hard to build up her fitness.

Being different

People can have very different heights. This is because height is controlled by several genes. Each gene has an effect on your height. Skin colour is also controlled by several genes.

A few features are controlled by just one gene. So there isn't as much difference between people. For example, you either have dimples, or you don't. You either have straight thumbs or curved thumbs.

More than one gene controls skin colour.

Dimples.

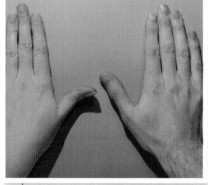

Curved or straight thumbs.

Two of each

Every human body cell has 23 pairs of chromosomes. You get one half of each pair from each parent.

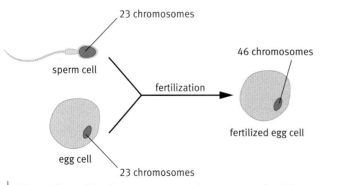

23 chromosomes

sperm cell

fertilization

46 chromosomes

fertilized egg cell

egg cell

23 chromosomes

The cells in this diagram are not drawn to scale. A human egg cell is 0.1 mm across. This is 20 times larger than a human sperm cell.

Each pair of chromosomes has genes for the same features. So you have two genes for every feature – one from your mum and one from your dad.

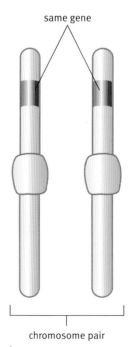

same gene

chromosome pair

These chromosomes are a pair.

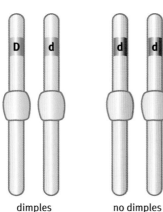

dimples   no dimples

Predicting genes

A gene can have different versions. For example, there are two versions of the dimples genes:

- for dimples = D
- no dimples = d

Darren doesn't have dimples. He has two d genes.

Maria has dimples. She has one D gene and one d gene.

The D gene is **dominant**. If you have a dominant gene it will always show up.

The d gene is **recessive**. Recessive genes only show up if you have two copies of them.

Darren and Maria decide to have children. There is a 50% chance that each child will have dimples.

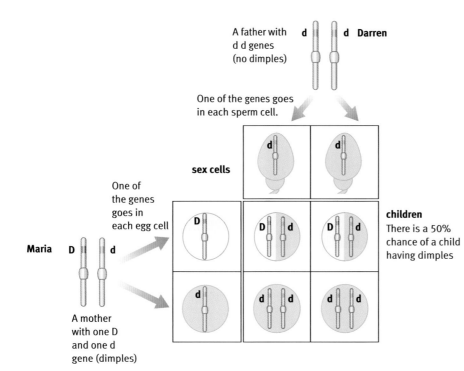

A father with d d genes (no dimples)   d   d   Darren

One of the genes goes in each sperm cell.

sex cells

One of the genes goes in each egg cell

Maria   D   d

A mother with one D and one d gene (dimples)

children
There is a 50% chance of a child having dimples

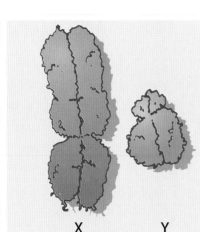

Diagram of a man's sex chromosomes.

X   Y

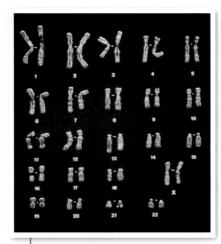

Chromosomes from a woman's cell.

Male or female?

Human body cells have 23 pairs of chromosomes. Pair 23 are the sex chromosomes. A woman's chromosomes look the same. They are both X chromosomes – XX. A man's sex chromosomes are different. He has one X chromosome and one Y chromosome – XY.

When genes go wrong

Genes are copied when new cells are made. Sometimes a mistake is made and the gene is changed. A few diseases are caused by faulty genes.

Your body makes mucus in your lungs and intestines. A faulty gene can make this mucus too thick. This causes the disease cystic fibrosis. People with cystic fibrosis get breathless and get lots of chest infections.

There are two versions of the mucus gene. One causes cells to make normal mucus. The other one causes sticky mucus. The normal gene is dominant. The faulty gene is recessive. A person must have two faulty genes to have cystic fibrosis.

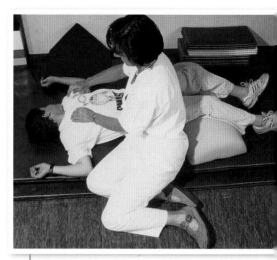

Tom has cystic fibrosis. He needs physiotherapy every day to clear his lungs.

Finding out about genes

Doctors can test a person's genes. They look for genes that can cause a disease. In a few cases doctors can check an embryo's genes. For example, they can find out if a baby is at risk of having cystic fibrosis.

If the embryo has inherited the disease the parents may decide to end the pregnancy. The mother has a termination (an abortion). This is a very serious decision. People have very different views about whether testing genes should be allowed.

Questions

1 Write a sentence about each of these words:

   nucleus chromosome gene dominant recessive

2 What do you think of these ideas?

   The genes you inherit from your parents control your life.

   No two people are exactly the same.

Microbes

Microbes are very small. They can only be seen under a microscope.

	Virus	Bacterium	Fungus
Size	20–300 nm	1–5 µm	50+ µm
Appearance			
Examples of diseases caused	flu, polio, common cold, AIDS, measles	tonsillitis, tuberculosis, plague, cystitis	athlete's foot, thrush, ringworm

Hundreds of thousands of bacteria can fit on a full stop.

A few kinds of microbes can make you ill.

For this to happen, the microbe must be passed on to you.
- A fungus causes ringworm. You have to touch an infected person to get the fungus.
- A bacterium causes plague. Rat fleas carry the bacterium.
- A virus causes a cold. The virus moves from person to person in tiny drops of water.

Natural barriers

There are microbes everywhere.

You have them all over your skin. They are in your mouth and nose.

You even have them in your gut – they make you fart!

Your body has ways of keeping out microbes that could make you ill.

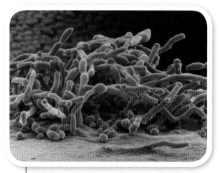

This fungus causes thrush. Thrush infects the mouth and the vagina (magnified about 800 times).

The average sneeze contains 2000–5000 droplets.

Chemicals in tears destroy microbes.

If microorganisms get in through your mouth, acid in the stomach destroys most of them.

Your skin produces chemicals that make it hard for microbes to grow.

The skin is a physical barrier to microbes.

Your body has ways of keeping out microbes.

Getting infected

Some microbes get past your outer defences.
They get into your body:

- if you cut yourself
- through openings in your body, like your mouth and nose.

Jolene cuts her finger. Conditions in her body are just right for microbes to grow. They have warmth, food, and water. The bacteria reproduce quickly.

It was just a small cut, so I ignored it. By the time I went to bed it was a bit sore and red. Now it's all swollen and shiny. It really hurts.

Fighting microbes

Jolene's body sends more blood to the area.

White blood cells surround the bacteria and digest them.

Jolene's body goes into action to fight the infection. More blood flows to her finger. It is carrying **white blood cells**. White blood cells are part of Jolene's **immune system**, which fights infections.

Many white blood cells die as they destroy the microbes. This leaves the wound full of dead bacteria and white blood cells – pus. The infection is over.

Immunity

When you get infected by a microbe you've not met before you may get ill. Your immune system fights off the microbe and you get better. Some of your white blood cells remember the microbe. If you are infected by this microbe again these white blood cells kill the microbe before it makes you ill. You are immune to the disease.

Vaccines

It takes your body a few days to fight off a microbe. Some microbes can be fatal in that time. **Vaccines** protect you from these microbes. A small amount of the microbe is injected into you. It is treated first so it cannot make you ill. Your white blood cells destroy the microbe and remember it. You are now immune from this microbe if you catch the real thing.

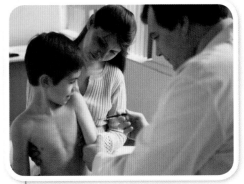

Vaccines are most important against infections that may be fatal.

Are vaccines safe?

Some parents worry about vaccinating their children. They hear on TV how some children react badly to a vaccination. The papers sometimes carry stories about people who may be damaged by vaccination.

Doctors decide that a treatment is safe to use when:
*   the risk of serious harmful effects is very small
*   the benefits outweigh any risks.

Vaccination is only effective if most people have it. Parents need to listen to what doctors say. They will explain how to weigh the benefits against the risks.

Cured

Jolene's doctor prescribed her some **antibiotics**. Antibiotics are chemicals that kill bacteria and fungi. They have no effect on viruses.

Resistance

Some bacteria are no longer killed by every antibiotic. They have become resistant to the antibiotic.

The diagram explains how this can happen.

Penicillin was the first antibiotic to be discovered. It saved many lives in World War II.

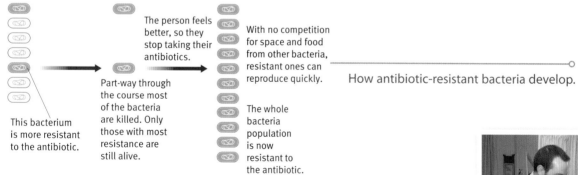

This bacterium is more resistant to the antibiotic.

Part-way through the course most of the bacteria are killed. Only those with most resistance are still alive.

The person feels better, so they stop taking their antibiotics.

With no competition for space and food from other bacteria, resistant ones can reproduce quickly.

The whole bacteria population is now resistant to the antibiotic.

How antibiotic-resistant bacteria develop.

You can help stop bacteria becoming resistant to antibiotics by:
*   always finishing a course of antibiotics so all the bacteria are killed
*   only using an antibiotic when you really need it.

Superbugs

Some bacteria are resistant to most antibiotics. They are called 'superbugs'. They aren't usually dangerous for healthy people. But if you are already ill and catch a superbug, then only one particular antibiotic can save you.

MRSA is a superbug that can infect people in hospital. Careful hygiene helps to stop it spreading.

Personal hygiene

Microbes can only harm you if they get into the wrong place. These rules of **hygiene** make it difficult for microbes to infect people:

- Wash and cover cuts to stop microbes getting through the skin.
- Avoid food that smells bad – it is probably full of microbes.
- Sneeze into a tissue to stop microbes getting to other people.
- Wash your hands after using the toilet to stop microbes from your gut getting to other people.
- Wash your hands before preparing and eating food.

Food hygiene

Warm raw food is a paradise for microbes. All the conditions are there to help them grow quickly. The poisons that microbes put into your food can make you ill. Here are a few rules of hygiene in the kitchen to stop your food from going off:

- Slow down the growth of microbes by keeping food cool.
- Keep food covered to stop insects transferring microbes to it.
- Wash knives and chopping boards in hot water to remove microbes from raw food.
- Cook food hot enough and for long enough to kill microbes.
- Keep raw meat separate in your fridge so microbes cannot pass to cooked food.

Keep raw food covered!

Questions

1 Write a sentence about each of these words:

microbe   white blood cell   immune system

vaccine   antibiotic   hygiene

2 What do you think of these ideas?

Doctors should only prescribe antibiotics as a last resort.

All children should be vaccinated against killer diseases.

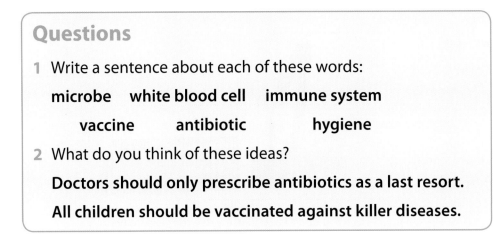

Some colours come from plants.

Plant dye

A dye is a chemical that can be used to colour cloth.

Many different dyes can be made from plants. Here is a method that works well with some plants:
- Grind up the plant material to increase its surface area.
- Boil it up with some water to break open the plant cells.
- Filter out the solid mush, leaving the dye dissolved in water.

Some dyes have a very interesting property. You can change their colour by adding acid or alkali to their solution. These dyes are **indicators**.

pH scale

Universal indicator is made from a mixture of different dyes.

It has lots of different colours, depending on its **pH**.
- Acids turn it yellow, orange, or red (pH 6 to pH 1).
- Alkalis turn it blue, indigo, or purple (pH 8 to pH 14).
- Pure water turns it green (pH 7).
- Pure water is **neutral**.

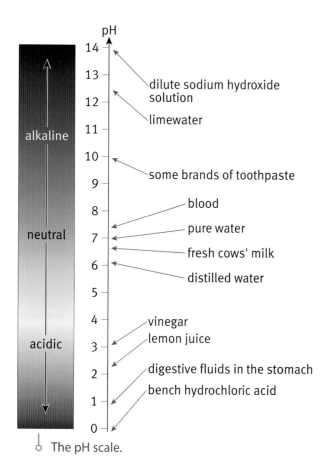

The pH scale.

pH
- 14 — dilute sodium hydroxide solution
- 13
- 12 — limewater
- 11
- 10 — some brands of toothpaste
- 9
- 8 — blood
- 7 — pure water
- 6 — fresh cows' milk
- 5 — distilled water
- 4 — vinegar
- 3 — lemon juice
- 2 — digestive fluids in the stomach
- 1 — bench hydrochloric acid
- 0

alkaline
neutral
acidic

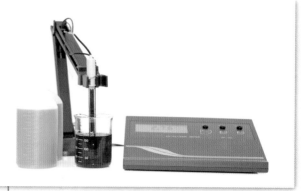

You don't have to use a dye to measure pH. This probe does it electronically.

Acids

There are many different **acids**. There are natural acids such as citric acid in lemon and lime juice and acetic acid in vinegar. There are laboratory acids such as hydrochloric acid.

Acids dissolve in water. Acid solutions turn litmus red. They have a pH below 7.

Acids can dissolve metals. Magnesium is a metal. The magnesium joins with part of the acid to make a salt. The reaction also makes hydrogen gas.

magnesium + hydrochloric acid ⟶ magnesium chloride + hydrogen

A mixture of hydrogen and oxygen explodes when it burns. In a test tube this just makes a 'pop'. This is the test for hydrogen. The hydrogen and oxygen join to make water.

Citric acid adds a sharp taste to food. It is the acid in lemon juice.

Alkalis

Alkalis are the opposite of acids. They are antacids. Alkaline solutions turn litmus blue. They have a pH above 7.

An acid is making a picture by dissolving parts of a metal plate. The rest of the plate is covered with wax so the acid cannot get at it. A feather brushes away the bubbles of hydrogen.

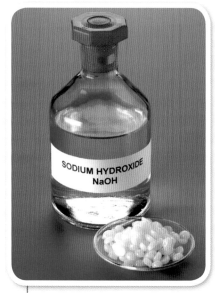

Pure sodium hydroxide is a white solid. It is soluble in water. It is an alkali. Sodium hydroxide is very corrosive.

Powerful alkalis such as sodium hydroxide are good at removing grease.

Alkalis neutralise acids. Calcium hydroxide neutralises hydrochloric acid to make calcium chloride (a salt) and water.

hydrochloric acid + calcium hydroxide ⟶ calcium chloride + water

Carbonates

Many rocks are made of carbonates. Marble is calcium carbonate.

Pour acid onto marble and it fizzes. It is releasing carbon dioxide gas.

Geologists use acid to test for carbonate rocks. They drip hydrochloric acid onto the rock. If it fizzes the rock is a carbonate.

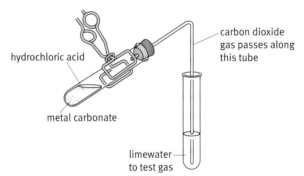

Testing for a carbonate rock with hydrochloric acid. This rock is limestone, made of calcium carbonate.

hydrochloric acid + calcium carbonate $\longrightarrow$ calcium chloride + carbon dioxide + water

hydrochloric acid

metal carbonate

carbon dioxide gas passes along this tube

limewater to test gas

The test for carbon dioxide is to bubble it through lime water. The lime water becomes cloudy and white when carbon dioxide bubbles through it.

Soil

Sam has a farm. He wants to grow beans next season. But he has a problem. Beans grow best if the soil pH is 6, slightly acidic. His soil has a pH of 5, too acidic for beans, and most other crops, to grow well.

So what does he do? He spreads lime on his fields to neutralise the acid in the soil. Lime is an alkali when it mixes with water. Sam has to be careful not to add too much lime. This would raise the pH above 7, which many plants don't like. It stops plant roots taking in important metals, such as iron and copper.

Spreading lime on a field can raise the soil pH.

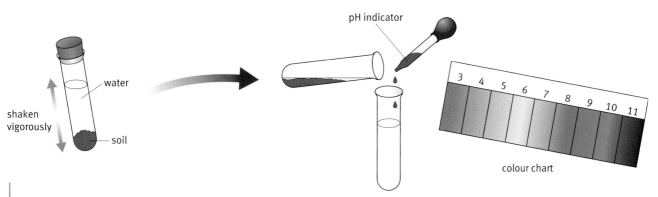

water

shaken vigorously

soil

pH indicator

colour chart

3 4 5 6 7 8 9 10 11

Sam has to test his soil regularly to check its pH.

Indigestion

Your stomach makes hydrochloric acid. The acid helps to digest food. It can be painful if the stomach makes too much acid. This can cause **indigestion**.

You can treat indigestion with antacids. The antacids neutralise the stomach acid. Many antacid brands contain a mixture of carbonates or hydroxides.

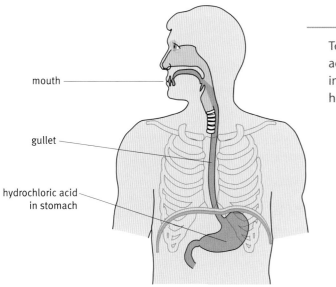

mouth

gullet

hydrochloric acid in stomach

Too much stomach acid can cause indigestion and heartburn.

Questions

1 Write a sentence about each of these words:

indicator pH neutral acid alkali indigestion

2 What do you think about these ideas?

Only natural dyes from plants should be used in clothes.

Acids are too dangerous to add to food.

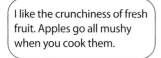

I like the crunchiness of fresh fruit. Apples go all mushy when you cook them.

Raw flour is horrible, but I like bread.

I really like the raw fish in sushi.

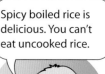
Spicy boiled rice is delicious. You can't eat uncooked rice.

Cooked or raw?

There are lots of different ways of cooking food:
- with a hot liquid or gas (such as water, steam, oil in a pan, or air in an oven)
- with heat from a red-hot grill
- with microwaves.

Why cook?

Cooking food raises its temperature. It changes food in different ways.
- Some foods get harder when cooked, some get softer.
- Some foods melt, others turn solid.
- Some foods take in water and swell up, others dry out and shrink.
- Many foods change colour.
- The taste nearly always changes.

Food is complicated stuff. It is often hard to explain the changes that happen when food cooks.

There are good reasons to cook food. These include:
- making the food feel nicer to eat (better texture)
- improving the taste
- providing meals that are easier to digest
- killing any microbes that are in the raw food.

A chef at work in a kitchen.

Hot potato

Raw potato tastes horrid and can give you stomach ache.

Cooked potato is quite different. It is soft, tastes nice, and is easy to digest.

Raising the temperature of a potato to 100 °C breaks apart the walls of the cells that store starch. The starch grains inside the cells soak up lots of water and burst open.

There are two reasons you can tell that cooking includes **chemical changes**:

* Chemicals in the food change into new substances.
* The food doesn't become raw again as it cools down.

Baking powder

Many chemical changes turn the ingredients into a cake as the mixture cooks.

Baking powder is a mixture of an acidic solid and a carbonate. Adding water and heating starts a chemical reaction. The reaction produces carbon dioxide.

A cake rises as it cooks. This happens because there is baking powder in the ingredients.

The baking powder reacts to make carbon dioxide gas as the cake cooks. The gas forms bubbles in the cake mixture. So the mixture rises. The cake mixture hardens at the end of cooking. This traps the bubbles of gas. The cake cannot collapse once it is hard.

Taking bread from an oven in a bakery.

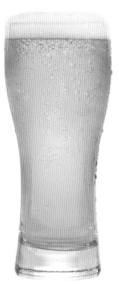
Beer is made by fermenting barley. The froth is just bubbles of carbon dioxide gas.

Soaps are made by boiling up a mixture of an animal fat or a vegetable oil with alkali.

Yeast

Bakers mix flour with water, fat, and yeast to make bread. The yeast is a microbe. It feeds on sugar to give alcohol and carbon dioxide. This is **fermentation**.

The sugar comes from starch in the flour. Yeast ferments best when it is warm.

The bubbles of carbon dioxide gas make the dough rise. Baking the dough traps the bubbles and evaporates the alcohol.

Yeast is also fermented in grape juice to make wine. This time the bubbles are allowed to escape, leaving the alcohol behind.

Washing up

Some food contains oil and fat. Oils and fats do not dissolve in water. This makes it hard to clean them off plates and dishes. Soap or detergent is needed.

Floating it free

Soaps and **detergents** break up the grease into small blobs and allow it to wash away in water. They are made up of tiny molecules.

The molecules have two ends. One end likes to be in water. The other end likes to be in grease. The parts that like grease allow the molecules to form a thin layer around any grease. This allows the grease to break up into tiny blobs. These can float freely in the water.

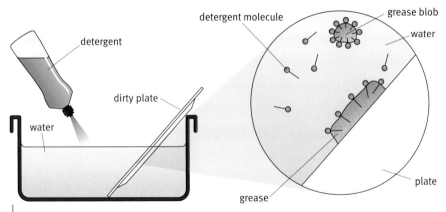
Detergents are synthetic. They are made from chemicals from crude oil.

Sticky stains

Food stains can be difficult to remove from clothes. They may contain proteins as well as grease. Then you need a biological washing powder. This acts in two stages:

- **Enzymes** digest the proteins, chopping them into small pieces until they are soluble in water.
- Detergent breaks the grease into small blobs that can mix with the water and float away.

Enzymes come from living things. They work best at body temperature. Some people have skin that reacts badly to the enzymes. They should not use biological washing powder.

Food stains may need more than just detergent.

Washing clothes

You can help a detergent by:

- shaking grease blobs off the cloth
- heating the water to soften the grease.

A washing machine does all this for you!

Care labels

Instruction	Symbol
Can use bleach	Cl
Wash at this temperature	60°C
Use a warm iron	
Use a cool iron	
Tumble dry	
Can dry clean	P
Don't iron	

Ignoring care labels can ruin your clothes.

Questions

1 Write a sentence about each of these words:

cooking chemical change fermentation soap detergent enzyme

2 What do you think of these ideas?

Cooked food is better for you.

Always look at the label before washing clothes.

Why paint?

Paint adds colour to our lives. We paint our homes to decorate them.

Paint also protects. It helps protect wood from rotting and metal from rusting.

Paint can also give a message. The lines on roads tell drivers where they can and cannot drive and park.

What is paint?

Brightening a room with paint. Emulsion paints are made using water. There is pigment in the paint. The water and pigment are mixed with a glue (binder) that sticks the pigment to the surface as the paint dries.

Painting a bridge in Newcastle on Tyne to stop it rusting. The pigment in this paint is white.

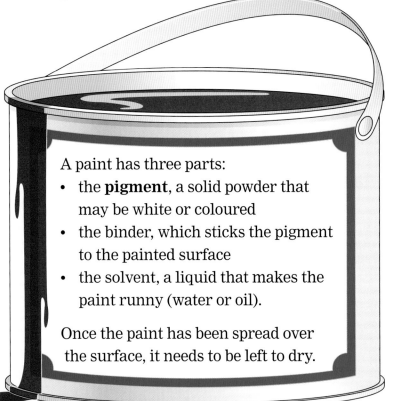

A paint has three parts:
- the **pigment**, a solid powder that may be white or coloured
- the binder, which sticks the pigment to the painted surface
- the solvent, a liquid that makes the paint runny (water or oil).

Once the paint has been spread over the surface, it needs to be left to dry.

Graffiti has a message.

Different paints

There are lots of different types of paint. Each one is useful in its own way.

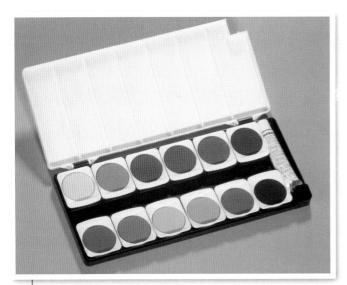

The pigment and binder come ready mixed in this paint box. Brushing water onto the colours makes paint when needed.

Oil paints are made with an oily solvent. The solvent evaporates as the paint dries. The binder slowly turns solid in air. This holds the pigment to the surface.

Some paints change colour as they get hotter or colder. The thermochromic paint on this mug changes colour when the mug contains a hot liquid.

Cleaning up

You have to use the right solvent to clean paint brushes. Watercolours use a binder that easily mixes with water. Brushes used with water colours can be cleaned with water. The binder makes a solution as it dissolves into the water. Hot water dissolves the paint quicker than cold water.

Oil-based gloss paint uses a binder that is **insoluble** in water. Washing the brushes in water is useless. You have to use the right solvent – white spirit.

White spirit is a solvent that can clean brushes after painting with an oil paint.

Heating lavender with steam distils off the perfume chemicals.

Harvesting lavender. Lavender is used to make perfume.

Why perfumes?

Perfumes make people and products attractive. Perfumes also help to hide nasty smells.

There are perfumes in cosmetics. There are also perfumes in many other products such as toilet cleaners, soaps, and air fresheners.

Natural perfumes

Some perfumes come from plants and animals. Rose oil comes from roses. Musk comes from a kind of deer that lives in China. A natural perfume may contain thousands of different chemicals.

Heating the plants with steam is one way to get the perfume. Another is to use a **solvent**. The perfume chemicals dissolve in the solvent to make a solution.

Synthetic perfumes

Some natural perfumes are very expensive. Chemists have found ways to make perfume chemicals. This can make them cheaper.

Natural fruit flavours are often mixtures of esters. It is quite easy to make esters from an alcohol and an acid such as acetic acid. Acetic acid is the weak acid in vinegar. Heating it with an alcohol and a drop of a strong acid quickly makes a fruity-smelling ester.

Lots of different **synthetic** perfumes can be made by mixing different esters. Not only are they cheaper than those from plants or animals but they can also make smells that can't be obtained naturally.

Cleaning a toilet. The liquid contains a detergent to clean the bowl, a disinfectant to kill germs, and a perfume for a nice smell.

Perfume properties

A perfume is a solution of smelly chemicals in a solvent. The perfume feels cool at first as the solvent evaporates straight away. This leaves a layer of perfume chemicals on the skin. Some of these chemicals **evaporate** quite soon. This is what you smell at first. An expensive perfume has many other chemicals that evaporate more slowly. This means that the smell changes and can get more attractive.

Safe cosmetics

Any perfume has to do more than just smell nice:

- It mustn't be poisonous.
- It mustn't react with skin
- It mustn't stain clothes.

All cosmetics are tested to see if they are safe. This includes perfumes. Testing has often involved animals. Most of the chemicals we use now have been tested on animals in the past.

Many people think that it is wrong to test cosmetics on animals. The European Parliament has now banned animal tests for cosmetics.

Perfume chemicals have a pleasant smell.

Testing mascara without using animals. This scientist is testing cosmetics on artificial skin.

Questions

1 Write a sentence about each of these words:

   paint   pigment

   insoluble   perfume

   solvent   synthetic

   evaporate

2 What do you think about these ideas?

   A life without paint would be very dull.

   Every chemical we use should be safety-tested.

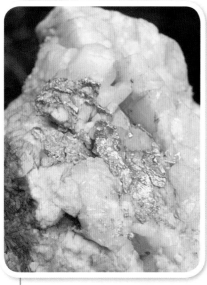

Gold in a rock. Gold doesn't react easily with anything else. It can be found as the free metal.

Panning for gold in Laos. The shiny gold is easy to spot.

Precious metals

Only a few metals are found in the ground in their pure free state. Examples are:

- gold
- silver
- platinum.

These metals are unreactive. They don't easily form compounds by reacting with air and water.

Panning

There is gold in some rivers. Gold is much heavier than the sand and mud. So the gold can be found by **panning**. The pan is used to scoop up some of the river bottom. Swirling water gently inside the pan moves the lighter sand and clay to the outside. The glinting gold stays in the middle – if there is any.

Jewellery

Gold, silver, and platinum are quite rare and they do not react with water or air. So they stay shiny. This makes them valuable and very useful for jewellery.

Nickel is a cheaper metal that is sometimes used in jewellery. It is quite common for people to be allergic to nickel. It causes skin rashes. A thin layer of expensive gold on a piece of nickel jewellery not only makes it look better – it can also stop people having an **allergic** reaction to it.

Metals in rocks

Most metals are found in rocks. Some rocks contain metal ores. Ores are compounds of metals with oxygen and other elements. It takes a chemical reaction to free a metal from its ore.

Making copper

Malachite is an ore of copper. The green rock turns black when hot. It turns into copper oxide.

Three steps can **extract** copper metal from malachite:
- crushing to reduce it to dust
- heating to break it down to copper oxide and carbon dioxide
- heating with carbon to take away the rest of the oxygen and leave the metal.

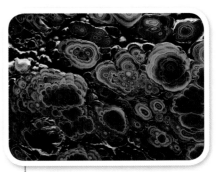

Malachite is a green rock. It contains copper, carbon, and oxygen. It is copper carbonate.

Using and re-using copper

There are miles and miles of copper wire in electrical goods. There is no better metal for electric wiring. But supplies of copper ore are limited – they could run out one day. Extracting and purifying copper from its ore is also expensive.

It makes sense to recycle metals after use. Recycling:
- means less mining and harm to the environment
- saves energy
- saves money.

Copper is the best metal for electrical wiring.

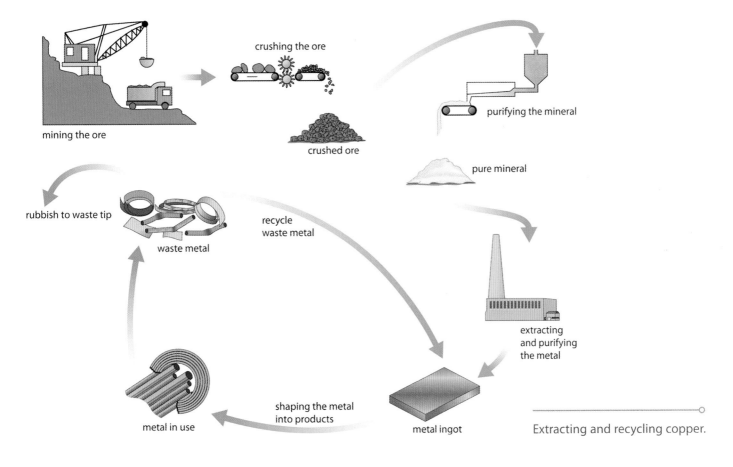

mining the ore

crushing the ore

crushed ore

purifying the mineral

pure mineral

rubbish to waste tip

recycle waste metal

waste metal

extracting and purifying the metal

metal in use

shaping the metal into products

metal ingot

Extracting and recycling copper.

Electroplating

Electricity is often used to put a thin layer of expensive shiny metal on top of a cheaper metal. Steel can be plated with shiny chromium.

Oscars are electroplated with four different finishes: copper, nickel, silver, and 24-carat gold.

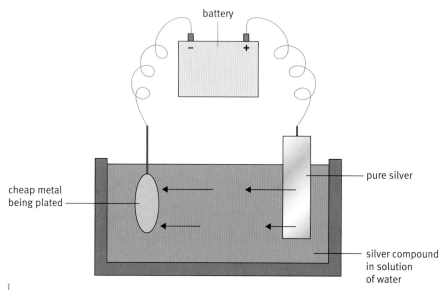

Using electricity to plate one metal onto another. This is **electroplating**.

Iron and aluminium

Iron and aluminium are much cheaper than metals used in jewellery.

Aluminium is much lighter than iron. It is less dense. It does not corrode. It is not magnetic.

Iron is strong. Iron girders hold up bridges and buildings. Strong iron makes cranes and cables to lift heavy weights.

Iron is heavy for its size. It is dense. It is also magnetic so it can be picked out of piles of rubbish with a big magnet.

Metals corroding

Iron rusts. It combines slowly with oxygen and water to swell into a weak brown crumbly solid. This is **rust**. Salt speeds up the process. So iron rusts faster near the sea.

Rusting is a form of corroding. Painting is one way of fighting against rust. It stops air and water getting to the iron. Keep the paint layer unbroken and the iron will last forever!

No rust

Aluminium doesn't rust. It always looks shiny. This is surprising, as the metal is much more reactive than iron. Any aluminium in air reacts with oxygen. A thin layer of hard aluminium oxide instantly forms over the surface. This stops any more air getting in. Aluminium makes its own protective coating.

Making cars

So what is the best metal for making cars? Aluminium or iron?

How about iron?
- stronger, so you need less metal
- cheaper to produce and shape
- high **density**, so makes the car heavier
- ends up as a heap of rust unless you paint it
- easy to separate with magnets for **recycling**

How about aluminium?
- doesn't corrode, so the car lasts longer
- weaker, so you need more metal or an expensive design
- difficult to join by welding
- doesn't have to be painted

An old chain turned rusty in air and water.

A Ferrari made from aluminium.

Questions

1 Write a sentence about each of these words:

panning   allergic   extract   electroplating

rust   density   recycling

2 What do you think about these ideas?

Plating a copper bracelet with gold is cheating.

We should be forced to recycle all of our metals.

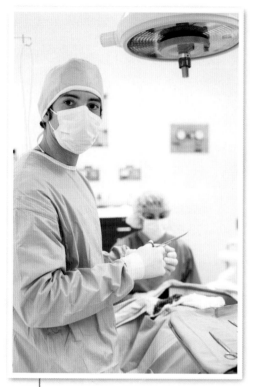

Clothes can tell other people what you do.

Garments

Clothes do different jobs such as helping us to keep warm and be safe. Without clothes you would be very exposed.

Fabrics

Clothes are made from **fabrics**. Fabrics are made from **fibres**. The fibres are woven or stuck together to make flat sheets of material. These are cut up into shape and fastened together to make clothes.

The behaviour of a fabric is fixed by:
*   the type of fibres in it
*   how those fibres are held together.

A warm jumper has loosely woven wool fibres. This allows it to be warm and floppy.

A shirt has polyester and cotton threads tightly woven together. The weave makes the fabric stiff, the mix of fibres make it cool and hard-wearing.

Natural or artificial?

There are two sorts of fibre.
*   **Natural** fibres come from plants or animals. Wool and cotton have short fibres. Silk has long smooth fibres.
*   **Artificial** fibres are made by reacting chemicals together to make a plastic. The plastic is squeezed out through tiny holes to make long and smooth fibres.

Clothes can keep you warm.

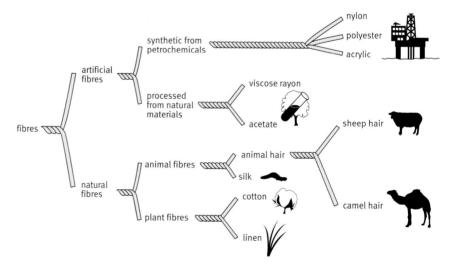

Fibres under the microscope

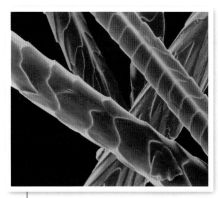

Wool fibres magnified 1000 times.

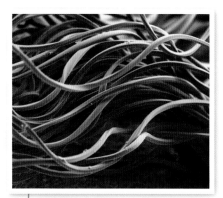

Polyester fibres magnified 75 times.

Cotton fibres magnified 600 times.

Stretch graphs

Testing a fibre tells you a lot about its properties. Weak fibres break at much smaller forces than strong ones. Stretchy fibres extend much more than stiff ones.

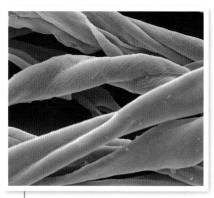

Nylon fibres magnified 100 times.

Why test?

John Fletcher is in charge of testing fibres for a company that makes sewing threads. He says:

'Quality control is very important for customer confidence. I use a machine to test threads chosen at random. Sometimes the stretch graph shows that the material isn't meeting the standard. I stop threads leaving the factory until the problem is sorted. It doesn't happen very often!'

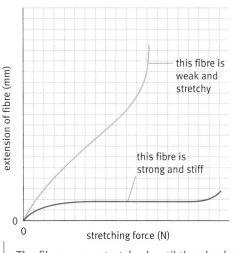

this fibre is weak and stretchy

this fibre is strong and stiff

extension of fibre (mm)

stretching force (N)

The fibres were stretched until they broke.

John Fletcher uses this machine to stretch threads until they break. It provides an extension–force graph for each thread.

Fibre	Useful property
wool	warm
lycra	stretchy
cotton	cool
polyester	hard-wearing

Mixing fibres

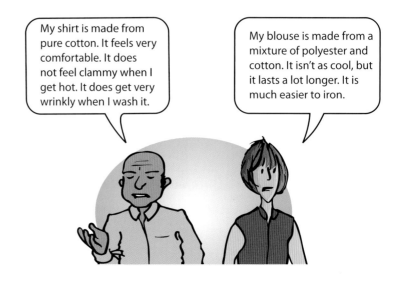

My shirt is made from pure cotton. It feels very comfortable. It does not feel clammy when I get hot. It does get very wrinkly when I wash it.

My blouse is made from a mixture of polyester and cotton. It isn't as cool, but it lasts a lot longer. It is much easier to iron.

Fishing in the North Sea. It is often cold and wet work.

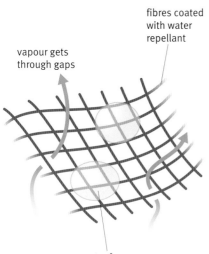

fibres coated with water repellant

vapour gets through gaps

water forms drops

Waterproofing

Getting wet and cold can be deadly. **Waterproof** clothing can save your life.

Fabrics have gaps between the fibres. These can let water through. Blocking them with wax or rubber can stop this. It also stops water vapour from your sweat escaping. You end up soaked in your own sweat.

Coating the fibres with a chemical that repels water is more effective. The gaps aren't blocked, so vapour can escape. Liquid water doesn't soak into the fibres and just falls off the breathable fabric.

Flameproofing

Fibres in **flameproof** materials have a chemical coating that makes it much more difficult to set the fabric alight. Furniture must be made with flameproof fabrics. Then it can't be accidentally set alight by cigarettes.

New fibres for old

Artificial fibres are better than natural ones for some purposes. These include:
- tents, which need to be waterproof
- sails, which need to be light and strong
- outdoor clothing, which needs to be breathable.

Stitching wounds

Fibres used for stitching wounds must have a special set of properties. They must be:

- strong, so that they don't break
- sterile, so that they can't infect the wound
- hypoallergenic, so that they don't react with the body
- flexible, so that they are easy to knot.

Some stitches can dissolve slowly, eventually disappearing completely. This is especially useful for surgery and for helping wounds to heal inside the body.

Covering wounds

Winston is a paramedic. He needs to use many different fibres and fabrics in his job.

'I usually use a sterile pad first,' says Winston. 'It can be pressed onto the wound to stop the bleeding. The cotton in it absorbs the blood.'

Sticking plaster holds the pad in place. The plaster needs to be stretchy.

Winston wraps elastic bandages tightly around swellings. 'They need a lot of artificial fibres to get the stretch into them,' says Winston. 'Pure cotton isn't much good.'

But cotton is ideal for supporting fractured limbs. It is flexible but strong, as well as feeling cool.

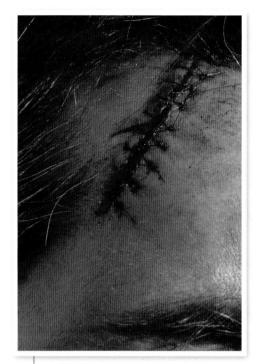

The sides of this wound are held together by stitches.

Winston uses many different fabrics.

Questions

1 Write a sentence about each of these words:

fabric   fibre

natural   artificial

waterproof

flameproof

2 What do you think of these ideas?

All clothes should be flameproof.

Natural fabrics are more comfortable than artificial ones.

The Earth from space. There are white clouds of water droplets in the atmosphere.

Gas	Typical percentage by volume
nitrogen (N_2)	78
oxygen (O_2)	19
water (H_2O)	2
argon (Ar)	1
carbon dioxide (CO_2)	0.04

The atmosphere

The air around you only goes up about 15 km into space. Above that there is almost nothing – certainly nothing you can breathe. You are adapted to thrive in an **atmosphere** that has about 20% oxygen and very little carbon dioxide.

Most of the atmosphere is nitrogen. It just sits there most of the time, not combining with anything.

Oxygen is very reactive, combining with almost anything if it gets hot enough.

The amount of water in the atmosphere varies, depending on when it last rained.

Argon just doesn't react with anything else.

That leaves carbon dioxide. Although there is very little, plants rely on it to make their food.

Pure and polluted air

Some diesel engines give out tiny specks of carbon in the exhaust. These black particles can cause lung damage. It is very hot inside engines. It is hot enough to form nitrogen oxide from nitrogen and oxygen in the air.

The air in city streets can be very polluted. The pollutants come from fuels burning in engines. Some cyclists wear breathing masks to protect themselves from particles that might harm their lungs.

Burning fuels

Burning fuels give the energy needed to generate electricity, to heat homes, and to drive vehicles.

The fuels that we burn are compounds of carbon and hydrogen.

- Burning carbon from a fuel combines with oxygen to make carbon dioxide.
- Burning hydrogen from a fuel combines with oxygen to make water.

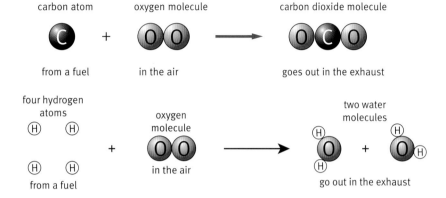

So burning a fuel can be summarised by this word equation:

fuel + oxygen $\longrightarrow$ carbon dioxide + water

Fuels and pollution

A **pollutant** is a chemical in your environment that is harmful.

There are three main pollutants from cars and lorries in the air you breathe:

- particles of carbon from engines that do not burn all the fuel
- nitrogen oxides – acid gases from car engine exhausts
- carbon dioxide – a greenhouse gas that traps heat.

The acid gases in polluted air can cause breathing problems. They also make **acid rain**. The greenhouse gases cause global warming. Their level is increasing year on year.

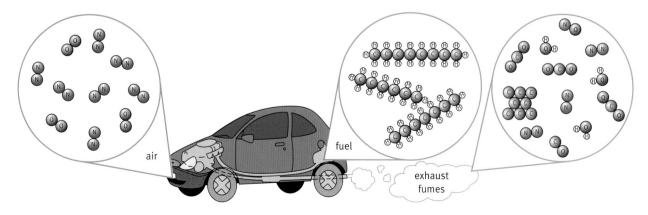

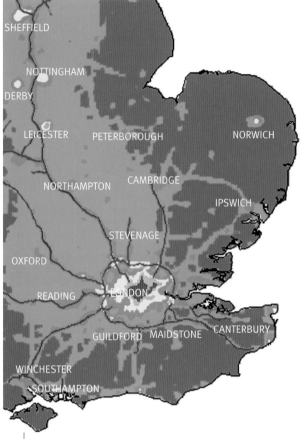

This map shows the concentration of pollution by nitrogen oxides over parts of England. The red lines show that it is highest in large cities and along motorways.

Air quality

Air pollution in the UK is measured automatically all the time. Robot sensors in key places transfer their data to central computers. Pollution levels are reported daily with some weather forecasts. On the Internet you can get hour-by-hour updates for where you live.

Acid rain

Rain isn't just water. It also contains small amounts of gases from the air. Pollutants can significantly lower the pH of rainwater, making acid rain.

- Sulfur dioxide from power stations dissolves in water to become sulfuric acid.
- Nitrogen oxides from car exhausts become nitric acid in water.

In some places the acid rain is often neutralised by the ground it falls on. Where this doesn't happen, acid rain can lower the pH of rivers and lakes. Eventually the plants die. Dead plants rot and fish die.

This equipment automatically measures air pollution day and night.

Acid rain has slowly eaten away this lion outside Leeds Town Hall. Erosion has taken away much of the detail.

Catalytic converters

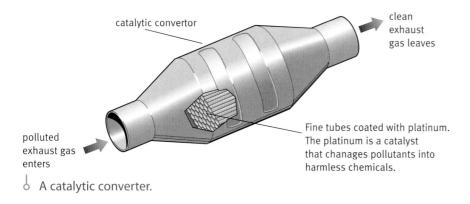

catalytic convertor

clean exhaust gas leaves

polluted exhaust gas enters

Fine tubes coated with platinum. The platinum is a catalyst that changes pollutants into harmless chemicals.

○ A catalytic converter.

Cars in the UK have their exhaust gases tested every year as part of their MOT test.

A **catalytic converter** in a car's exhaust system can remove pollutants. It changes the pollutants into harmless gases.

Two features of the converter allow it to do this:
- The platinum coats the surface of the tubes in the honeycomb.
- The thousands of tiny tubes give a large surface area for the gases to react.

The platinum is a catalyst. It speeds up the reaction between the exhaust gases. The catalyst changes nitrogen oxides into nitrogen, which is harmless.

No pollution please

We know how to remove pollutants, but each process has a price tag.

A catalytic converter adds to the cost of a car. The converter also lowers the kilometres per litre. The driver has to pay for more fuel.

Catalytic converters work with petrol engines but not with diesel engines. They do not remove the fine particles of carbon from buses and lorries.

Electric trams, cars, and trains help to cut air pollution in a city. Most of the electricity they use comes from power stations outside the city. Power stations produce pollutants when they burn fossil fuels.

Questions

1 Write a sentence about each of these words:

atmosphere   burning

nitrogen oxides

acid rain   pollutant catalytic converter

2 What do you think about these ideas?

People who cause pollution should pay to clean it up.

Electric vehicles are better than petrol or diesel cars in towns.

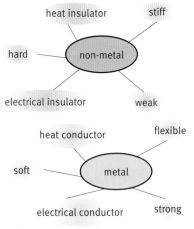

Typical properties of metals and non-metals.

Metals

Your world is made of stuff. That stuff can either be a **metal** or a non metal.

Is it cold to touch? Is it strong and flexible? Does it conduct electricity? Is it ductile (soft enough to be drawn into wires)?

If the answer is yes to these questions, then it's probably a metal.

Useful alloys

Many useful things are made from mixtures of metals. These are **alloys**.

Electronic components are connected by solder. Solder is an alloy of lead and tin that melts easily.

This musical instrument is made from brass. Brass is an alloy of copper and zinc.

All the aluminium you see is likely to be an alloy with another metal such as copper (duralumin). Aluminium alloys are light for their size and used to make things like aircraft parts. Pure aluminium is not strong enough.

The longest steel-arch bridge in the world. The steel in this bridge is a strong alloy of iron and carbon.

Some strong magnets are made from an alloy of aluminium, cobalt, and nickel called alnico. These magnets are used in electric motors and loudspeakers.

Smart alloy

Nitinol is an alloy of nickel and titanium. The alloy is tough like most metals. You can twist or stretch it out of shape without snapping it. But if you put it in hot water it suddenly goes back to its shape.

Filling teeth

I'll drill out the rot first. Then fill the hole with an alloy of mercury.

Isn't mercury poisonous?

It sets hard quickly, so the mercury can't escape.

How long will it last?

Years. The alloy expands as it goes hard, so it fits the hole tightly.

Pure metals

Pure metals are often quite useless.

Coins made out of pure copper would be too soft, so coins are made out of an alloy of copper and other metals. This makes them harder.

Pure copper is best for carrying electricity but it isn't very strong, so electrical connectors often use brass instead. This alloy of copper and zinc doesn't conduct electricity quite as well but it is much stronger than pure copper.

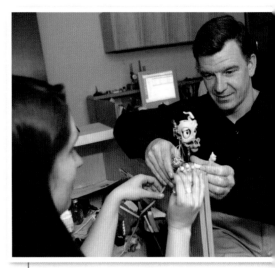

Alloy wires make the toy move when they get hot.

Electrical plugs use brass connectors.

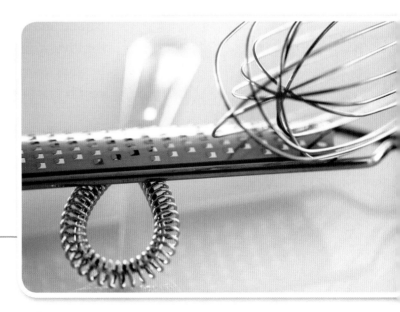

Ordinary steel goes rusty in air and water. Stainless steel stays bright even in a busy kitchen.

Hardness	Mineral
10	diamond
9	corundum
8	topaz
7	quartz
6	feldspar
5	apatite
4	fluorite
3	calcite
2	gypsum
1	talc

A scale of hardness.

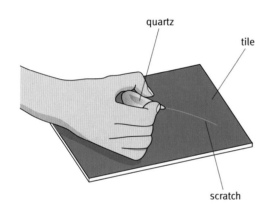
quartz
tile
scratch

Amethyst is made of the same chemical as sand. It is purple because it includes small amounts of impurities.

Minerals

Rocks are made of many chemical compounds. Each compound is a **mineral**.

Some rocks contain mostly one mineral. Other rocks contain a mixture of minerals.

Sometimes the minerals in rocks form crystals. If they are hard enough, these can be worn as jewels. The colour of a jewel often comes from small amounts of impurity in the mineral.

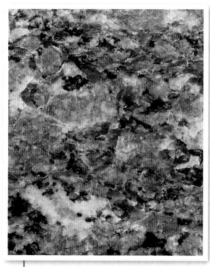
Granite is made from many crystals of different minerals.

Marble is made from just one mineral, calcium carbonate.

Bricks

Bricks are made from clay. Clay contains several minerals as tiny crystals. Once water gets between the crystals, the clay can be shaped easily. Firing the clay at high temperature removes the water. Then some of the crystals in the clay melt to stick everything tightly together as the brick cools. The result is a very hard material known as a ceramic.

How hard?

Sarah tests the **hardness** of ceramic tiles. She scratches them with a set of standard minerals. 'I start with the softest, talc, and work up to the hardest, diamond,' she says. 'The first one to make a mark on the tile tells me how hard it is.'

One tile isn't marked until she gets to quartz. Sarah gives it a hardness of 6.5, between quartz and feldspar.

Natural rock

Bricks are strong and hard. That is why they are used to make buildings. Many natural materials are just as good, or better. Granite, limestone, and marble are rocks that are often used in buildings. They come straight from the ground but are expensive to cut into shape. This means that they cost more than bricks but can look better and last longer.

Marble floor tiles look good and are hard-wearing. They are also expensive.

Artificial rock

Concrete is made from cement, sand, gravel, and water. It is widely used in construction. Here are some reasons why:

- It starts off liquid, so it can be poured into any shape.
- It sets solid over a few days, even under water.
- It is hard, strong, and cheap.

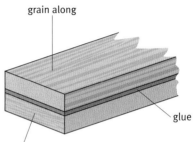

Plywood is a composite material.

Composites

Like all rock, **concrete** is hard and strong when it is squashed. It isn't tough, so it cracks easily when it is stretched. Reinforcing the concrete with rods of steel gives a material that combines the properties of rock and metal.

Plywood is another **composite** material. It is made from thin sheets of wood, bonded together with glue. Wood is only strong along its grain. Bonding the sheets together gives a material that is strong in all directions. Plywood is also cheap and light but does rot when wet.

Glass-reinforced plastic (GRP) is made from mats of glass fibre. These are soaked with a resin that slowly sets to make a solid polymer. GRP combines the strength of glass with the flexibility of plastic.

A modern vaulting pole is very advanced. It is made of a composite with layers of GRP and carbon fibre.

Questions

1 Write a sentence about each of these words:

metal   alloy   smart alloy

mineral   hardness

concrete   composite

2 What do you think of these ideas?

Metals are only useful when mixed with each other.

New materials are spoiling sport.

Earthquakes shake the ground suddenly and violently.

Ash and rock from volcanoes destroy local plants and wildlife. People may die if buried by ash falls.

Tsunamis are sudden huge waves from the sea. They can cause widespread flooding and damage.

Grape vines at the foot of an active volcano. Volcanic soils are very fertile. This is why many people live near them.

Volcano

Volcanoes can erupt without warning. Red-hot lava pours down the side of the volcano. It cools and becomes hard rock. Massive amounts of ash billow out, filling the air with dust. As the ash falls on the ground it smothers plants and animals. They don't have a chance.

Earthquake!

Below ground, rocks can be under great pressure. Sometimes the rocks move suddenly, releasing a lot of energy. This is an **earthquake**. The ground shakes where the shockwave reaches the surface. Buildings can collapse, dams give way, bridges fail. The damage can be immense.

Earthquakes under the sea set up huge waves called **tsunamis**. These sweep inland with no warning, causing great destruction.

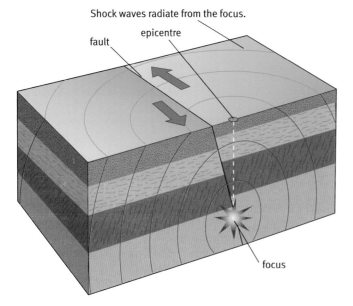

Shock waves radiate from the focus.

fault

epicentre

focus

Damage limitation

So how can governments protect people from these natural disasters? A government can:

- educate people so that they understand the risks
- evacuate them if they know a natural disaster is going to happen
- set up an early warning system to detect it
- make plans for dealing with it
- have regular practice drills.

Moving continents

The idea that continents move was first suggested by Alfred Wegener in 1915. He noticed that he could fit the continents together like a jigsaw. The types of rock matched when the continents were put together. He also found that the same fossils appeared in more than one continent, in bands that also fitted the jigsaw.

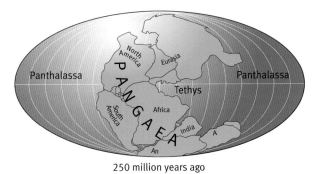

250 million years ago

Wegener showed how all the continents could once have formed a single continent.

Many scientists didn't accept Wegener's theory.
- It was impossible to show that the continents were moving.
- They couldn't figure out what could be pushing the continents around.
- They had another theory to explain how mountains and oceans formed as the Earth cooled and got smaller.

Wegener's theory is now accepted.
- Satellite measurements can show the continents moving.
- The theory of plate tectonics can explain how the continents move.
- The cooling-Earth theory does not work.

The cooling-Earth theory. As the Earth cools down, it shrinks like a rotting apple. The highest ridges become mountains.

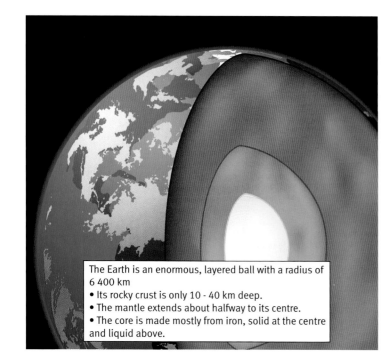

The Earth is an enormous, layered ball with a radius of 6 400 km
• Its rocky crust is only 10 - 40 km deep.
• The mantle extends about halfway to its centre.
• The core is made mostly from iron, solid at the centre and liquid above.

The new theory

The Earth is less solid than you think.

We live on the thin outer layer called the crust. It is cool and solid, made from many different compounds. This floats on a thicker layer of hotter plastic rock called the mantle. Beneath this is the hot liquid outer core made of iron. At the centre is the solid inner core of iron.

Cracks in the crust

Earthquakes happen at the edges of the great slabs of rock that make up the Earth's crust. These slabs are called **tectonic plates**. The rigid plates move very slowly. In some places they push against each other. In other places they pull apart.

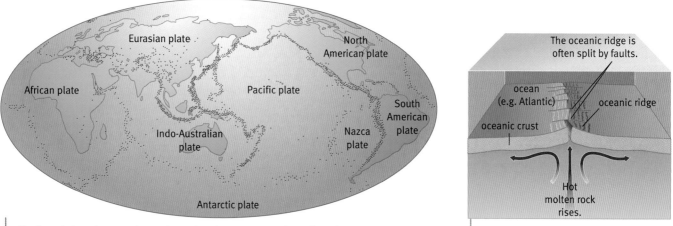

Each red dot shows where there has been an earthquake. There are more earthquakes in some places.

New rock is made where plates move apart from each other.

Rocks from magma

Volcanoes form where magma from the mantle pushes its way through cracks in the crust. **Magma** is molten rock from underground.

When magma reaches the surface, it is called **lava**. Lava flows out of volcanoes as a river of liquid rock, cooling to form rock.

Magma on the surface cools fast. This makes an **igneous rock** called basalt. The crystals in this rock are very small. Sometimes molten rock cools deep underground. Then the magma cools very slowly to make rocks such as granite. The crystals in these rocks are quite big. You will often see crystals in granite where polished rock is used to decorate buildings.

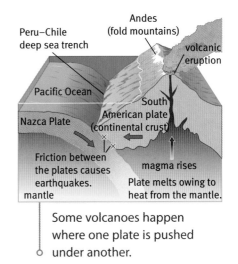

Some volcanoes happen where one plate is pushed under another.

Polished granite from Scotland. This igneous rock cooled slowly deep in the Earth. It is used to decorate buildings.

These hills in India formed about 65 million years ago. There were many volcanic eruptions. Gradually layer upon layer of basalt created mountains of new rock.

Questions

1 Write a sentence about each of these words.:
**volcano   earthquake   tsunami
tectonic plate   magma   lava
igneous rocks**

2 What do you think of these ideas?

One day scientists will be able to predict earthquakes.

Nobody should be allowed to live near a volcano.

Gunpowder exploding.

It can take a long time for an old car to rust away.

Fish on ice to keep them fresh.

Fast or slow?

Some **chemical reactions** are very fast.

- Gunpowder reacts in an instant.
- Forest fires burn fiercely as the wood combines with oxygen in the air.

Other reactions are really slow.

- Cars take many years to rust through completely.
- Food takes a few days to go rotten.
- Acid rain dissolves stone buildings very slowly.

The hotter the faster

Many reactions are slow when cold. They need a hot flame or spark to get them started.

Once started, some reactions give out a lot of energy. The reaction gets faster as the **temperature** goes up. This happens during a fire.

Cooling things is a good way to slow down a reaction. Pouring water onto a fire helps to cool it. This can slow it down so much that it goes out.

A fire blazes faster as it gets hotter and hotter. Water cools burning wood and helps to slow down the reaction.

Keeping food

Making food colder slows down all the chemical changes that make food go bad. Fish shops display their food on ice. Your refrigerator helps you to preserve food. Putting food in a freezer slows down the rotting even more.

More or less?

Things burn much more rapidly in pure oxygen (100%) than in air (20% oxygen). This proved disastrous during the first Apollo mission to the moon. The space capsule was filled with pure oxygen during a practice before the launch. There were three astronauts on board. A fire started and spread very fast. All three were killed.

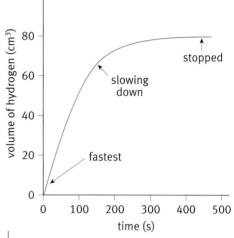

A graph showing the volume of gas collected in a syringe during an experiment. The graph is steep where the reaction is fast. The reaction stops when all the marble is used up.

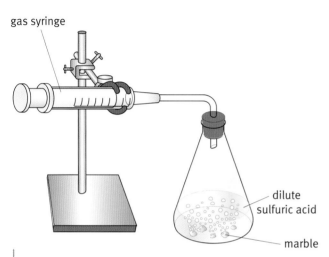

The gas syringe measures the amount of gas given off by the marble as it reacts with the acid.

Increasing the **concentration** of chemicals speeds up a reaction. Concentrated acid solutions react faster with metals than dilute solutions.

Many useful reactions involve chemicals dissolved in water. As the chemical gets used up, its concentration goes down. This slows down the reaction. The reaction stops when one of the chemicals is all used up.

A chemical reaction happens when there are collisions between the molecules. If the conditions are right, they combine to make new particles. The speed at which this happens depends on the concentration.

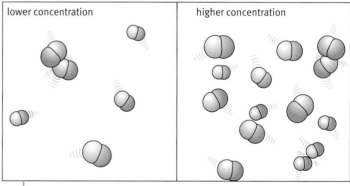

Increasing the concentration gets the particles closer to each other. Putting particles closer together means that they collide more often. The reaction goes faster.

Too much acid in the stomach can be painful.

The small bits burn faster than the large ones. The fire burns faster as it gets hotter.

Large or small?

Jane has indigestion. She decides to take a tablet to ease the pain.

The instructions tell her to chew the tablet before swallowing it.

Chewing the tablet grinds it into a powder. This allows it to react more quickly with the acid in her stomach.

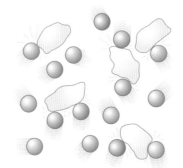 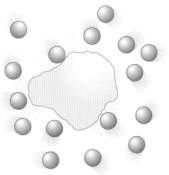

several small lumps   one big lump

Lots of small lumps have more surface area than one big lump. The reaction goes faster because there are more places where the chemicals can react.

A liquid can only react with a solid on its surface. The bigger the **surface area** of the solid, the faster the reaction. Here are some examples:

- Sliced bread goes stale faster than a whole loaf.
- Small twigs burn much faster than big logs.
- Flour dust in air can explode, but wheat grains are no danger.

Small droplets of fuel can explode in air.

Catalyst control

'I measure pollution in the exhaust,' says Jake. 'If there's too much, I have to fail the car.'

Jake lets the car warm up before he does the test. The platinum **catalyst** in the exhaust system works better when it's hot. Reactions on the surface of the platinum convert the pollutants into harmless gases.

'The platinum in the converter should last forever,' says Jake. 'It doesn't get used in the reactions, just helps them along.'

Platinum is very expensive, so it is spread in a thin layer inside the converter, in thin tubes. This increases the surface area for the reactions to take place.

Jake carries out MOT tests on cars.

Natural catalysts

Enzymes are natural catalysts. Enzymes help to digest food by speeding up chemical reactions. Enzymes in saliva turn starch into sugar. You can taste this as you chew bread.

Enzymes in your stomach and intestines break down proteins to make them soluble. The soluble chemicals can pass into your blood. Manufacturers add the same type of natural enzyme to washing powder to break down protein stains on clothes. This makes soluble chemicals that easily wash away.

There is an enzyme in living cells that breaks down a harmful chemical called hydrogen peroxide. The enzyme is catalase. It helps hydrogen peroxide break down into water and oxygen.

Catalase from liver speeds up the breakdown of hydrogen peroxide. A little detergent in the solution means that the mixture froths up as oxygen gas forms.

Questions

1 Write a sentence about each of these words:

   chemical reaction   temperature   concentration

   surface area   catalyst

2 What do you think of these ideas?

   Cooling is a better way of keeping food fresh than using preservatives.

   I could not live without enzymes.

Unmixing

A sieve separates grains of rice from the rest of the plant mixed up in the harvest.

A large bag acts as a filter in this street cleaner. It separates the dust and dirt from the air.

Sewage flows slowly through this tank to allow solids to sink to the bottom.

Oily dirt is hard to separate from cloth.

Clothes after dry cleaning. A special liquid dissolves and washes away the greasy dirt.

Does it dissolve?

Oil and grease do not dissolve in water so many clothes are washed in dry cleaning liquid. This dissolves the grease to make a **solution**. The fabric does not dissolve. A rinse with clean liquid gets the dirty solution out of the clothes. Warming evaporates any remaining liquid to leave the clothes clean and dry.

Cheshire salt

Water is used to dissolve things on a big scale. The chemical industry uses lots of water to get the salt from underground in Cheshire. Pumps force water down pipes into the salt. The salt dissolves.

The salty water comes back to the surface. Filters remove any sand and rock. Then heating evaporates the water to leave pure salt.

So dissolving and **filtering** can be used to separate a chemical that is soluble from chemicals that are insoluble.

Successful separations

Sand is a useful raw material. A lot of it is used to make concrete. Sometimes sand comes mixed up with a lot of salt. Salt is not good for concrete so the sand is washed before it is used.

- The sand–salt mixture is added to water.
- The salt dissolves in the water, but the sand doesn't.
- The sand is allowed to settle.
- The salt solution is poured away, leaving the sand.

The process is repeated with fresh water until all of the salt has gone.

Pouring off the water from the sand is an example of **decanting**.

Concrete starts off as a mixture of sand, cement, gravel, and water.

Chromatography

Mixtures of substances with different colours can be separated by **chromatography**. In this way scientists can spot forgeries or check that the dyes used to colour food are safe.

This is how you do it:

- Dissolve the mixture of colours in a liquid.
- Put a spot of the solution on a start line near one end of a piece of special paper.
- Dip the paper in another liquid in a jar.
- Wait for the liquid to move up the paper.
- Dry out the paper.

The colours move up the paper with the liquid. The colours move at different speeds. Colours that stick to the paper move slowly. Colours that are more soluble in the liquid move faster.

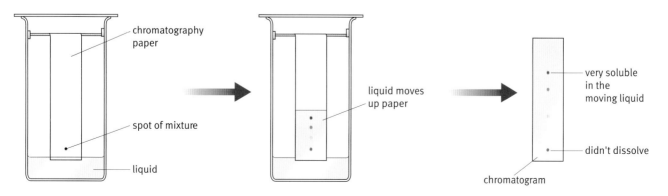

chromatography paper

spot of mixture

liquid

liquid moves up paper

very soluble in the moving liquid

didn't dissolve

chromatogram

Scrapyard separation

Sam runs a scrapyard. He makes money from recycling items made from metal as metals have different properties.

'I always carry a magnet,' says Sam. 'Iron is the only metal it sticks to.'

Copper is valuable. Its brown colour makes it stand out amongst the other silvery metals. Gold is another metal with a special colour, but Sam has never found it in his yard.

Sam picks up a heavy piece of grey metal. 'Lead is easy. It's heavy and soft.'

Aluminium is just as easy because it's so light and shiny.

A magnet separates iron from other metals in this scrapyard.

Magnetic materials

Iron and steel are magnetic. You use them in lots of different ways.

- in compasses to help you find your way
- in electric motors for CD players and food mixers
- in fridge doors to hold them shut.

Medical separations

Sorting things out is important in medicine. Separation techniques are used in diagnosis and in treatment.

Blood is a mixture of cells and a liquid called plasma. Spinning blood in a **centrifuge** separates the blood cells. The heavy blood cells end up at the bottom, so that the lighter plasma can be decanted off.

Your liver breaks down waste products in your body. It turns some wastes into urea. Your kidneys separate the poisonous urea from your blood.

People whose kidneys don't work have to use a dialysis machine. Their blood goes past a thin membrane, full of tiny holes. Small urea molecules pass through. The large cells stay in the blood.

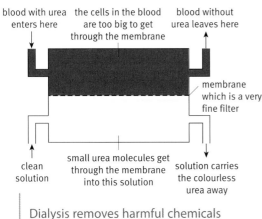

blood with urea enters here

the cells in the blood are too big to get through the membrane

blood without urea leaves here

membrane which is a very fine filter

clean solution

small urea molecules get through the membrane into this solution

solution carries the colourless urea away

Dialysis removes harmful chemicals from blood.

Fresh water from the sea

The sea is mainly salt dissolved in water. The water can be separated from the salt by **distillation**.

Pure water normally boils at 100 °C. Seawater boils at a slightly higher temperature. The water boils off the salt–water mixture. The salt stays behind.

The hot water vapour passes through a condenser to cool it down. The vapour condenses to form liquid water.

Pure ice

When sea water gets cold enough, the water turns into ice. Water normally freezes at 0 °C, but salt makes it freeze at a lower temperature.

The ice is pure. The salt is left in the liquid sea water.

Strong drink

Alcoholic drinks, such as wine or beer, are made by allowing a fungus called yeast to ferment in a mixture of water and sugar. If there is no oxygen, the yeast converts the sugar into alcohol and carbon dioxide.

The yeast dies if the alcohol concentration is too high.

Some drinks, such as gin and whisky, have alcohol concentrations of 40% or more. They are made by distillation. Alcohol boils at a much lower temperature than water, so it evaporates first when the mixture is heated.

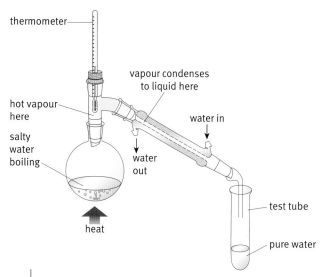

Distilling salty water to get pure water.

This iceberg is solid pure water floating in a salt–water mixture.

Questions

1 Write a sentence about each of these words:

solution   filtering   decanting

chromatography   centrifuge

distillation

2 What do you think about these ideas?

Dry cleaning is better than washing.

There would be nothing worth drinking without filtering and distillation.

Leaving evidence

I'm doing two years for burglary.

I left a footprint behind on the floor.

Well, it's obvious. They'd leave evidence behind at the scene.

I thought of that. I wore the socks on my hands - but I still got busted!

How did they catch you?

That was silly. Why didn't you wear shoes?

So would your fingerprints.

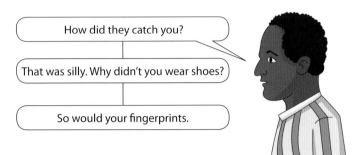

The crime-scene investigator is wearing a paper suit, overshoes, and gloves, to make sure the evidence is not contaminated.

Collecting evidence

Crime-scene investigators look out for **evidence** that could identify the criminal.

- Are there any fingerprints?
- Did the criminal leave bits of their clothes behind?
- Are there any footprints that could identify their shoes?
- Is there any blood at the scene that could be used for DNA matching?

Evidence is always photographed before it is removed. Then it is put in a plastic bag. Each bag must be:

- sealed, to avoid contamination
- labelled, so that it can be clearly identified.

This attention to detail is important. The evidence may be used in court to convict a criminal.

Taking fingerprints

Your skin is covered all over with oils. You leave these on every surface you touch. Dusting a metal powder on the surface makes these oils show up. The powder sticks to the oil, showing all the ridges on the skin.

Any fingerprints found are photographed. Then they are lifted onto sticky tape to be kept as evidence

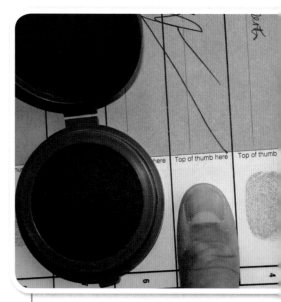

Police often take fingerprints to eliminate people from their enquiries.

A forensics officer uses tape to lift fingerprints from a gun found at the scene of a crime. He is wearing protective clothing to prevent contamination of the scene.

Matching fingerprints

The pattern of ridges on a finger is unique. No other finger has the same pattern of loops, whorls, and arches. Even identical twins have different fingerprints!

If your fingerprint is found at a crime scene, then you must have been there.

loop   whorl   arch

Three basic features can be used to describe all fingerprints.

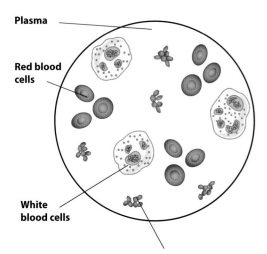

Plasma

Red blood cells

White blood cells

Platelets

Blood

Traces of blood are often found at a crime scene. Blood has four main parts:

- red blood cells, which carry oxygen
- white blood cells, which protect against infection
- platelets, which allow blood to clot
- plasma, which is mostly water.

Blood contains DNA, which can sometimes be used to match a blood sample to its owner.

Blood groups

Your red blood cells belong to one of four groups. They can be A, B, AB, or O. You inherit your **blood group** from your parents.

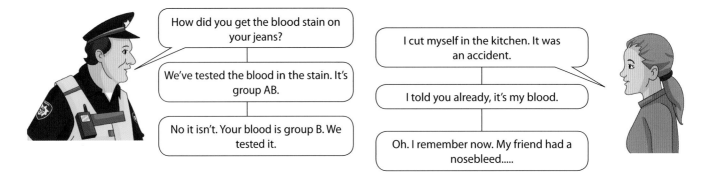

How did you get the blood stain on your jeans?

We've tested the blood in the stain. It's group AB.

No it isn't. Your blood is group B. We tested it.

I cut myself in the kitchen. It was an accident.

I told you already, it's my blood.

Oh. I remember now. My friend had a nosebleed.....

Taking a sample from blood-stained jeans.

DNA

Every cell in your body contains the same DNA. Your genetic code is held in your DNA. Everybody has a different genetic code.

Even the tiniest scrap of skin, hair, or blood from a crime scene can contain DNA. Scientists can chop the DNA into bits and spread them out to make a DNA profile. If two DNA profiles are the same, then the two samples probably belong to the same person.

Forgery

Many crimes involve pens. Signatures are easily forged. Getting hold of the owner's pen is not so easy. Scientists can use chromatography to compare samples of ink.

- A spot of each ink is placed at one end of the paper.
- That end is left in a solvent that dissolves the dyes in each ink.
- The solvent climbs up the paper, separating the dyes in the inks.

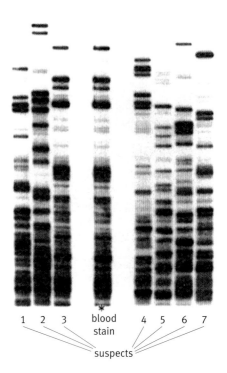

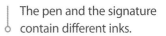

1   2   3   blood   4   5   6   7
            stain
            suspects

Did this pen really sign the cheque?

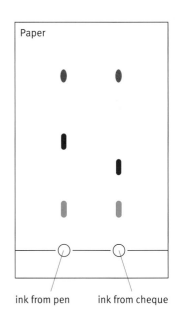

Paper

ink from pen   ink from cheque

The pen and the signature contain different inks.

Questions

1 Write a sentence about each of these words:

evidence   fingerprints   blood group

DNA profile   chromatography

2 What do you think about these ideas?

The police should have a copy of everyone's DNA profile.

You can't commit a crime without leaving some evidence at the scene.

An oil refinery. The tall columns are distillation columns. They separate the chemicals in crude oil.

Crude oil

Crude oil is a sticky black fluid. It was formed millions of years ago from dead sea creatures.

Oil is trapped by rocks underground. Some oil comes from deep below ground under the sea. At an oil well the oil flows up through a long pipe.

Crude oil is a mixture of many different chemicals. Most of them are made from two elements: hydrogen and carbon.

Distilling fractions

An oil refinery separates the chemicals in oil.

The chemicals are separated by heating.

Some oil chemicals evaporate more easily. They rise higher up the distillation tower.

Other chemicals are much harder to evaporate. They collect towards the bottom of the tower.

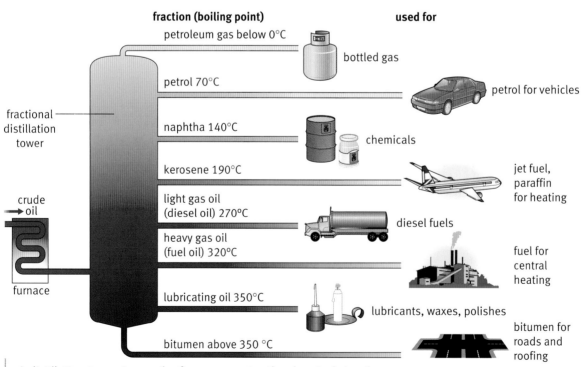

A distillation tower in an oil refinery separates the chemicals in oil.

Fractions

Each fraction of crude oil is made from carbon and hydrogen. It is called a **hydrocarbon**.

The carbon is in the form of long chains with the hydrogen stuck on the sides. Each fraction has molecules of similar length.

The low-boiling-point fractions have short molecules.

The high-boiling-point ones have long molecules.

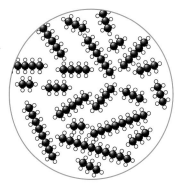

Crude oil is a mixture of hundreds of different hydrocarbons.

Plastics

Many plastics are made from crude oil. The short hydrocarbons are made into short monomers.

These can then join end to end to make very long polymers. This process is called polymerisation.

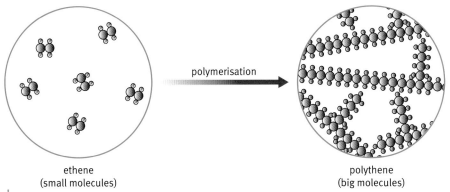

ethene
(small molecules)

polymerisation

polythene
(big molecules)

When polythene is made, the atoms in the small ethene molecules are rearranged to form big polythene molecules.

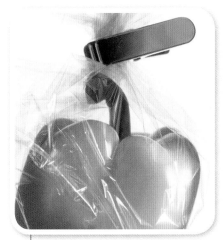

Polythene bags help people to protect, store, and carry food.

Chemicals that burn

The gases from oil burn very easily. They light if you just put a match to them. Petrol is another fule that burns easily.

Diesel is much harder to light. Diesel fuel needs more heating to get it to burn.

A Bunsen flame is blue and clean if the air hole is open.

This gas isn't mixed with enough air. It burns with a sooty yellow flame.

Carbon monoxide has no smell. It is colourless. It is very poisonous and so a detector can save lives.

Burning

Fuels can burn without smoke if there is plenty of air. Hot fuels react rapidly with oxygen.

Burning gives out lots of energy. The flame is very hot.

The reaction makes new chemicals. These are waste products. They include:

- carbon dioxide
- water (steam)
- carbon (soot)
- carbon monoxide.

Carbon and **carbon monoxide** only form if there isn't enough oxygen for complete combustion. To get a hot blue flame from burning gas you need to mix it with the right amount of air. Then it produces only carbon dioxide and water when you light it.

Incomplete combustion is a bad thing.
- The flame gives out less energy.
- The carbon forms black soot, which pollutes.
- Carbon monoxide is a colourless gas with no smell. It is very poisonous.

Safe heating

Ted services gas central heating systems.

One of the first things he checks is the colour of the flame.

'A yellow flame means not enough air is getting to it,' says Ted. 'Where there is soot, there is also carbon monoxide. Soot is messy, but carbon monoxide is a killer.'

Ted can turn the system off if there isn't enough fresh air getting into the house.

'It's for their own good,' he says. 'They won't notice the carbon monoxide building up. They'll just fall unconscious and it can be fatal.'

Petrol or diesel

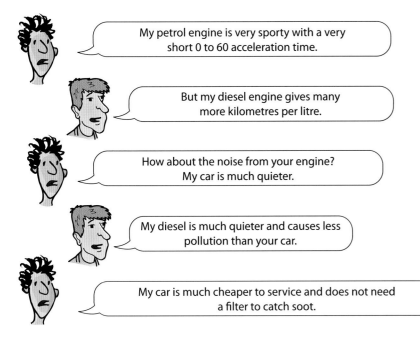

My petrol engine is very sporty with a very short 0 to 60 acceleration time.

But my diesel engine gives many more kilometres per litre.

How about the noise from your engine? My car is much quieter.

My diesel is much quieter and causes less pollution than your car.

My car is much cheaper to service and does not need a filter to catch soot.

Fuel choice

Gas is the main fuel for heating in many homes. Gas arrives as you need it, along a pipe. It is a clean fuel. The main disadvantage of gas is that leaks can lead to explosions.

Oil has some disadvantages:

- It is not such a clean fuel.
- You have to keep it in a large tank near the house.
- You need to order fresh deliveries when the level runs low.

Of course, there is the cost of the fuel itself. You don't want to pay more than you have to.

If you live a long way from a gas supply your heating fuel may have to come by road.

Questions

1 Write a sentence about each of these words:

    crude oil   petrol   diesel
    fuel   burning   carbon monoxide

2 What do you think of these ideas?

    Crude oil is too precious to burn.
    Petrol engines are better than diesel engines in cars.

Additives

Colour is important for some foods.

There are several good reasons why manufacturers add small amounts of chemicals to our food. Here are some of the things they add:

- preservatives, to stop mould and bacteria growing
- antioxidants, to stop oxygen in the air reacting with the food
- colourings, to make the food attractive or interesting
- flavourings, to add taste
- enhancers, to make the taste come out more.

These **additives** help the manufacturers to make food you want to buy.

But they can't add just anything. It has to be tested. It has to be safe.

Preservatives

microorganisms cannot grow in concentrated sugar

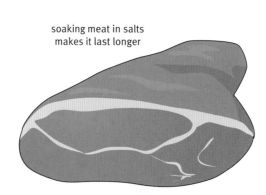

soaking meat in salts makes it last longer

preserved ham

ONIONS

onions vinegar

vinegar is too acid for microorganisms to grow

Apples in the air

Jean is making an apple tart. 'I always wipe the cut surfaces with some lemon juice,' says Jean. 'It stops them going brown.'

Each time Jean slices through the apple she breaks lots of its cells. Chemicals burst out of the cells and quickly react with oxygen in the air. With some apples this makes an ugly brown colour.

Lemon juice contains vitamin C. This is an **antioxidant**. It reacts with oxygen in the air, preventing it from getting to the apple.

Salt

The salty taste in your food comes from a chemical called sodium chloride. You can add more salt to make the food taste better.

It also acts as a preservative. Enough salt stops the food from rotting.

Salt for your food comes the sea or under the ground. Evaporating sea water leaves salt crystals behind.

Water is pumped into underground salt deposits. The salt dissolves in the water. The salty water is then pumped to the surface where it is allowed to evaporate.

This can sometimes leave large holes deep in the ground. Sometimes the holes collapse, changing the shape of the ground above.

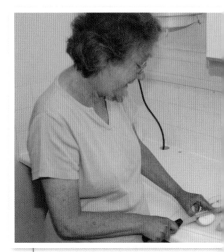

Some slices of apple turn brown quickly.

Salt cod in a fish market in Barcelona.

Salt pans near Mahabalipuram in India. Seawater runs into the pans and evaporates in the hot sunshine. Salt crystals are left behind.

Mining rock salt.

Subsidence caused by salt mining in 1891.

E numbers

This is full of additives with E numbers. It can't be good for me.

If it's got an E number, then it's probably safe.

But they aren't natural. Look at this one – Lecithin (E 322). Where's that from?

It's an emulsifier from soya beans. Not artificial at all.

Many additives have **E numbers**.

This means that they pass the European Union safety tests.

Both natural and artificial additives have E numbers. Only flavourings don't have E numbers.

E number	Type of additive
E100 to E199	colour
E200 to E299	preservative
E300 to E399	antioxidants
E400 and above	emulsifiers and others

Some additives are now banned because scientists have shown that they are not safe. Here are two examples:
* a food colouring that can make children behave badly
* a preservative that increases the risk of cancer.

Food labels

Some people are allergic to particular additives. They become ill if they eat them in their food.

They need to look carefully at the labels on the food they buy.

The label should show everything that is added to the food during its manufacture.

This drink has lots of additives.

Vitamins

You need small amounts of **vitamins** in your diet to keep healthy. Your body cannot store some vitamins so you need to eat them every day. Some foods have added vitamins to make sure you eat enough. The recommended daily amount (**RDA**) of a vitamin shows how much you need.

Vitamin	Examples of why you need it
A	allows your eyes to work properly
B	helps you to get energy from your food
C	allows your body to make the tissue that holds together your muscles and bones
D	helps you to absorb calcium from your food and form strong bones

Vitamin E. Some people take extra vitamins to make sure they get the recommended daily amount (RDA).

Sweeteners

Many people like food that tastes sweet. This may be because natural **sweeteners**, like sugar, have lots of energy. Artificial sweeteners are used in diet drinks instead of sugar. Tiny amounts of an artifical sweetener can replace a lot of sugar in a low-calorie food.

Deadly diet

Too much sugar in your diet may be bad for you. If you don't use all the energy in your food, you store it as fat. People who are overweight are much more likely to get heart disease.

Questions

1 Write a sentence about each of these words:

   additive   antioxidant   E number

   vitamin   RDA   sweetener

2 What do you think about these ideas?

   Food would be boring without additives.

   Additives with E numbers are safe.

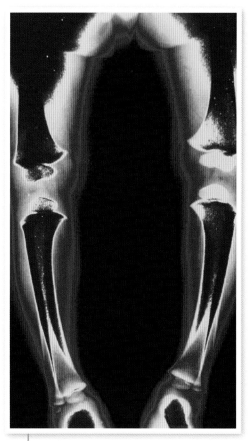

X-ray showing the bones in the legs of a child with rickets. This child did not get enough vitamin D in their diet. Without vitamin D the bones are not hard enough, so they bend.

Texting uses microwaves to carry a message away from the phone.

Messaging

Every message needs something to carry it – a medium.

Paper is the medium for a handwritten message. Air is the medium for a spoken message.

Walkie-talkies use radio waves. Fixed phones use copper wires.

Signal, range, and noise

A **signal** carries the message you are trying to communicate. **Noise** is meaningless data that gets picked up in a communication.

Noise can introduce errors into the message received.

Secure code

Many messages are sent in **code**.

Code can make a message clearer. Saying *"Alpha Bravo Charlie"* comes across clearly, whereas "ABC" might not.

Code can also be confusing. Anyone with a receiver can listen to radio communications. The message only makes sense if you know the code.

Morse code

Morse code uses dots and dashes to transmit letters of the alphabet.

It works with any signal that can be turned on and off:
* electricity in wires
* light in an optical fibre
* radio waves through the air.

Morse code is understood all over the world. Radio amateurs and people with a variety of physical disabilities use it to communicate.

Quick communication

Many communication systems use waves:
* Wireless networks use radio waves.
* Mobile phones use microwaves.
* Remote controls use infrared waves.

All three waves travel at the speed of light, which is 300 000 kilometres per second in space, so the messages also travel at 300 000 kilometres per second!

Code words	Meaning
ten-one	weak signal
ten-two	good signal
ten-three	switch off
ten-four	yes
ten-five	pass on
ten-six	busy
ten-seven	can't help
ten-nine	repeat
ten-ten	no
ten-twelve	stand by

Some radio 10 code words.

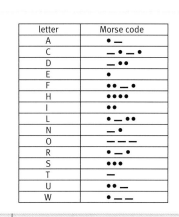

letter	Morse code
A	• —
C	— • — •
D	— ••
E	•
F	•• — •
H	••••
I	••
L	• — ••
N	— •
O	— — —
R	• — •
S	•••
T	—
U	•• —
W	• — —

Morse code for the most used letters.

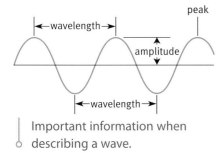

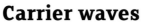

Important information when describing a wave.

Carrier waves

These features of a wave allow it to carry information:

- amplitude, which is how tall its peaks are
- wavelength, which is the distance from one peak to the next
- frequency, which is the number of peaks passing a point in one second.

Analogue or digital

There are two sorts of information: analogue and digital.

- **Analogue** has lots of different values.
- **Digital** has only fixed values – often just 'on' and 'off'.

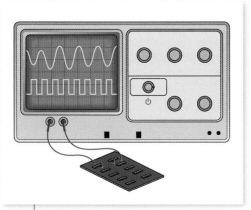

The display shows analogue and digital signals from an electronic circuit.

An indicator light is a digital signal.

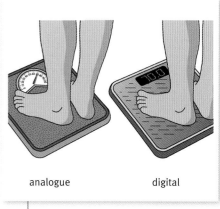

analogue   digital

Mass can be displayed as analogue or digital information. This person's mass is 70 kilograms.

AM radio broadcasts and analogue TV use analogue signals. Both have the problem of noise interfering with the signal. DAB radio and digital TV use digital signals. They can usually get rid of noise, so that the signal comes through clearly.

Wireless technology

Sam uses a laptop computer. The office has a wireless network. Wherever he is in the building, he can exchange information with any other computer. The radio signals are too weak outside, so he loses connection. When he works at home, he plugs into the office through the Internet. 'Working on the end of a wire is slow,' says Sam, 'and I have to stay at the desk'.

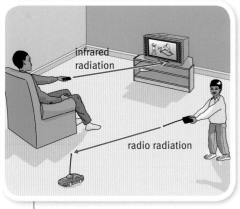

Radio and infrared travel at the same speed.

Are phones fatal?

Mobile phones and microwave ovens both use microwaves. Water is good at changing microwave energy into heat.

Ovens heat up food, so they have a power of about 900 W. Phones have a much smaller power of 0.25 W. This is enough for the signal to reach a nearby phone mast. Although the microwaves heat the water in your brain, the effect is tiny.

A phone mast has a larger power, about 100 W. However, the strength of the microwaves drops rapidly with distance.

How safe is this?

Risk reduction

Over 50 million people use mobile phones in the UK. So far, there is no evidence that they have any harmful effects. But if you are worried you can:

- use a hands-free kit
- only send text messages.

These keep the phone away from your head. You can also:

- keep your calls brief
- switch off your phone between calls.

All of these reduce the microwave power to your brain.

Mobile masts

Phone masts are necessary to relay signals to distant mobile phone users. The best place to put a mobile phone mast is:

- in a location where people want to use mobile phones
- high up, to increase its range
- where people can't get too close.

Governments set legal limits to the radiation from mobile phone masts.

Questions

1 Write a sentence about each of these words:

 signal   noise   code   analogue   digital

2 What do you think of these ideas?

 Mobile phones make people safer.

 Digital devices work no better than analogue ones.

Portable equipment uses batteries.

Battery

Jim's torch needs a **battery** of three cells.

Each cells supplies 1.5 V. Three cells in line gives him 4.5 V.

This is enough to make the lamp shine brightly.

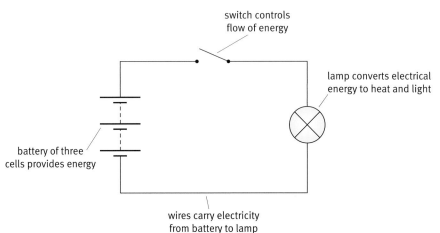

switch controls flow of energy

lamp converts electrical energy to heat and light

battery of three cells provides energy

wires carry electricity from battery to lamp

Which battery?

Batteries come in all shapes and sizes.

Zinc–carbon batteries are the cheapest. Alkaline cells last longer, but also have to be thrown away when they run out of energy.

Nickel–cadmium batteries are the most expensive of all, but they are rechargeable so they can be used over and over again.

This lead–acid battery is large, so it can store lots of energy. It is also rechargeable.

Chemical cells

All cells are made the same way. A conducting solution separates two electrodes made from different materials.

Chemical reactions take place at each of the electrodes.

This makes electricity flow through circuits connected across these terminals (electrodes). One electrode becomes positive, the other becomes neagative.

The cell is dead when the chemicals are used up.

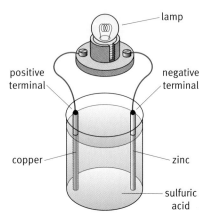

lamp

positive terminal

negative terminal

copper

zinc

sulfuric acid

This lithium cell doesn't store much energy. It can only be used once.

Fossil fuels

Just four different fuels are used to make most of our electricity:

- coal
- natural gas
- crude oil
- uranium

These are all dug out of the ground. The first three are **fossil fuels**. They are the remains of living things that died a very long time ago. Coal is made from dead plants. Crude oil and natural gas are made from dead sea creatures.

Fossil fuels won't last for ever. Some may run out in your lifetime. Once they have been used, they can not be replaced for millions of years.

Burning fossil fuels also increases the amount of carbon dioxide in the atmosphere. This is a greenhouse gas. It is a cause of global warming, making the Earth hotter.

Most UK electricity is generated by burning fossil fuels.

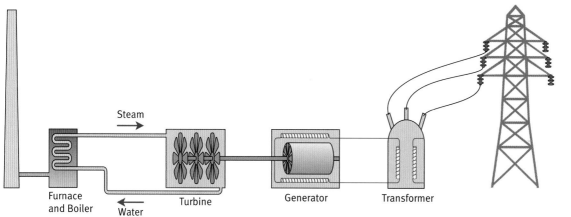

Making electricity

A **power station** makes electricity in four stages:
- Fuel burns to release heat.
- Water boils to make high-pressure steam.
- Steam forces the turbine to turn the generator.
- The generator makes electricity.

Energy lost

A power station produces heat by burning fuel. Not all of that heat is used to make electricity.

Heat escapes:
- in the exhaust gases from the burning fuel
- as the steam cools
- from the hot surfaces of the boiler and turbine.

Power grid

Each power station feeds its electricity into the National Grid. A step-up transformer raises the voltage of the electricity from the power station. The voltage can be as high as 400 000 V. The **power grid** carries electricity all over the UK. Step-down transformers reduce the electricity to 230 V for consumers. The high voltage reduces the amount of energy lost to heating the cables, so saves energy.

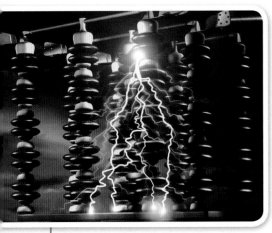

High-voltage cables are dangerous. The electricity can easily jump through the air to nearby objects.

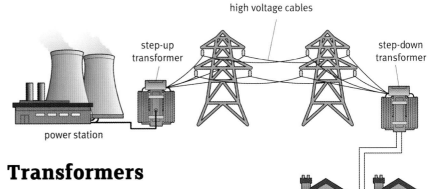

Transformers

A transformer has three parts:
- a primary coil of wire
- a loop of iron called a core
- a secondary coil of wire.

The primary coil takes alternating current (a.c.) electricity from the power supply. This magnetises the iron core. The secondary coil uses the magnetism to generate electricity at a different voltage.

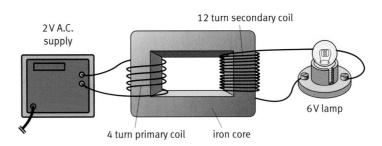

Payback times

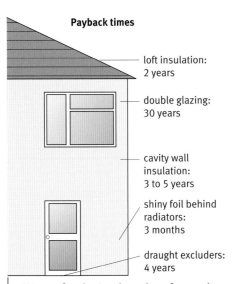

loft insulation: 2 years

double glazing: 30 years

cavity wall insulation: 3 to 5 years

shiny foil behind radiators: 3 months

draught excluders: 4 years

Ways of reducing heat loss from a house.

Saving money

Electricity costs money. You can use less electricity at home by:
- switching off the lights when you leave the room
- turning off devices that are on standby
- using high-efficiency appliances
- turning down the heating thermostat.

Power plates

Think before you plug a new appliance into the mains!
- Does it match the 230 V UK supply voltage?
- Can it cope with the 50 Hz mains frequency?
- Will the socket handle the power?

Switching on the appliance costs money. You have to pay for the electricity it uses.

How much you pay depends on:
- the power of the device
- how long you leave it switched on.

You have to pay for each **unit** of electricity it uses – even when it is on stand-by!

EASY HEAT
230 V 50 Hz
2.2 kW
DO NOT WET

Fan heater power 3 kW
Operating time 6 hours
Electricity used = 3 × 6
= 18 units

Multiply the power of an appliance with the time it is switched on to find how many units of electricity it uses.

Metering electricity

We need to check this electricity bill.

The present reading is 62161 units. Take away the previous reading of 61367 units.

That gives 794 units. What's the unit cost?

It says here 8.25p per kWh. A unit is the same as a kWh.

794 × 8.25 gives £65.50.

That's what it says here. We need to switch things off more.

The meter reading tells your supplier how many units you have used.

Questions

1 Write a sentence about each of these words:

battery   fossil fuel   power station

power grid   unit

2 What do you think of these ideas?

People should only be allowed to use rechargeable batteries.

Leaving things on stand-by wastes valuable fuels.

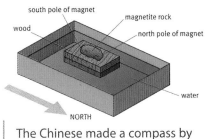

The Chinese made a compass by floating magnetite rock on water.

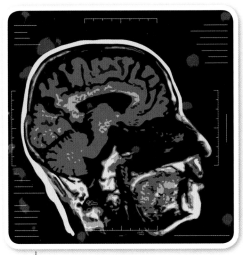

A very strong magnet is needed to make this MRI scan of a human head.

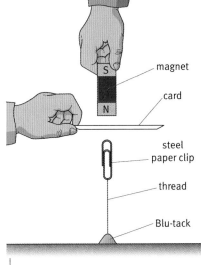

The magnet pulls up the steel paperclip through the card!

Swinging magnets

The Chinese probably discovered magnets first.

They found a special type of rock (called magnetite). When it floats in a dish of water it points north. Always.

All magnets try to line up in a north–south direction.

The north pole of the magnet points north.

The other end of the magnet is called its south pole.

Repel or attract

Magnets push and pull against each other.

Poles that are the same repel (push against each other).

Different poles attract (pull towards each other).

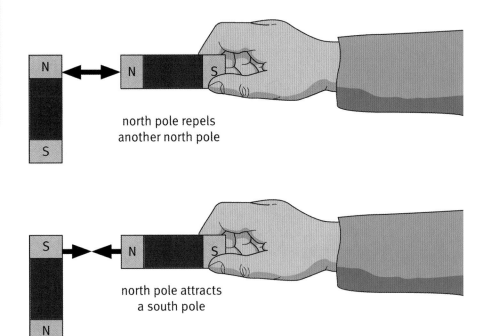

north pole repels another north pole

north pole attracts a south pole

Is it magnetic?

Very few materials are magnetic. Magnetism goes straight through all non-metals and most metals.

Iron and steel are **magnetic materials**. They are attracted to both poles of a magnet.

Only a magnet can repel another magnet.

Magnetic field

Every magnet is surrounded by an invisible force field, called a **magnetic field**. This allows it to exert forces on other magnets and magnetic materials placed nearby.

A compass can be used to make a useful picture of a magnetic field.

The arrows drawn on each field line for a bar magnet go from the N pole to the S pole.

A compass needle shows the direction of a field line.

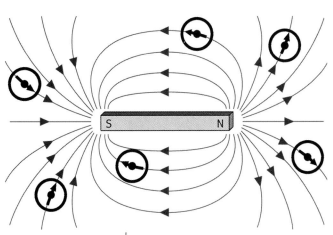

A compass shows the magnetic field of a bar magnet.

Compass

The magnet in a **compass** always lines up north–south.

This is how to find out your direction:
* Hold the compass level in front of you.
* Rotate the case until the N mark is at the red end of the needle.
* Find the direction you want by reading the compass scale.

You have to watch out for large iron and steel objects nearby. They can distort the Earth's magnetic field.

Most of this compass is made of non-magnetic materials. The steel magnet swings freely on an aluminium pivot.

Fields by filings

Iron filings line up along magnetic field lines.

This can be used to get a quick picture of a magnetic field.
* Put the magnet under a piece of card.
* Sprinkle iron filings over the card.

The filings are pulled to places where the field is strong.

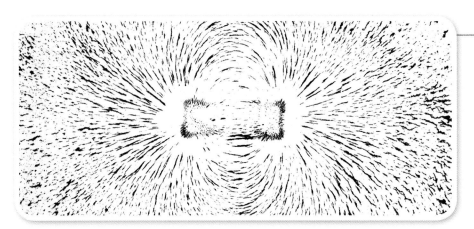

Iron filings show the magnetic field of a bar magnet.

The Sun's violent surface throws out stuff called cosmic rays.

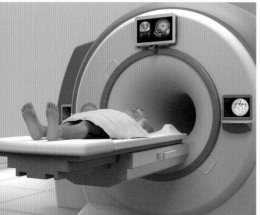

This display of light in the sky comes from cosmic rays entering the Earth's atmosphere.

MRI scanners contain strong electromagnets.

Earth's field

Cosmic rays are bad news.

They are fast-moving particles from space. Many of them come from the Sun.

If they hit one of your cells, they can damage its DNA, and may eventually cause cancer.

Fortunately, a lot of them get trapped by the Earth's magnetic field. They spiral round the field lines and enter the atmosphere near the North or South Pole.

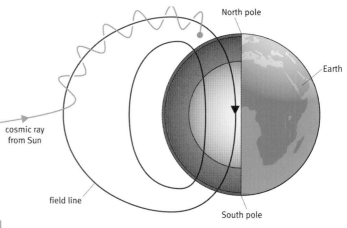

Cosmic rays from the Sun are guided to the Earth's poles by its magnetic field.

Electrical magnetism

You can magnetise some things by making electricity flow around them.

As shown below, the electricity flows through insulated wire. The wire is coiled around a magnetic material, such as iron or steel.

Iron loses its magnetism when the electricity stops. Steel stays permanently magnetised.

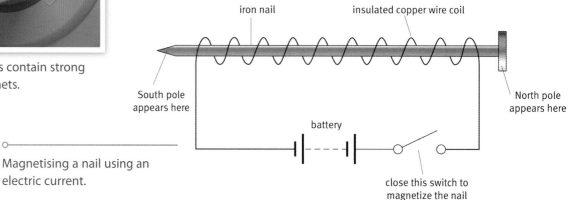

Magnetising a nail using an electric current.

Electromagnet

Useful **electromagnets** have several important features:

- lots of wire coiled around the core
- a core with lots of iron in it
- a large current in the wire.

Of course, they can only be used for picking up things made of iron or steel.

Loudspeaker

Loudspeakers have three parts:

- a coil of copper wire that acts as an electromagnet
- a permanent magnet made of steel
- a paper cone to move the air.

The electromagnet is fastened to the paper cone.

Alternating current (a.c.) in the coil makes the cone vibrate, making the sound.

Copying a magnet

You don't have to use electricity to magnetise a piece of steel.

Stroking it with a magnet works just as well.

But you need to make sure that you always stroke the same way!

Questions

1 Write a sentence about each of these words:

   magnetic material   magnetic field

   compass cosmic ray   electromagnet

2 What do you think about these ideas?

   GPS has made the magnetic compass redundant.

   The Earth's magnetism is good for my health.

This electromagnet is switched on to pick up scrap metal. It lets go of the scrap metal when it is switched off.

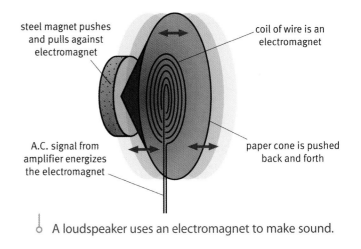

steel magnet pushes and pulls against electromagnet

coil of wire is an electromagnet

A.C. signal from amplifier energizes the electromagnet

paper cone is pushed back and forth

A loudspeaker uses an electromagnet to make sound.

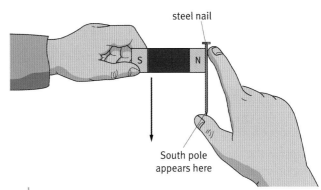

steel nail

South pole appears here

Magnetising a nail by stroking it with a magnet.

Changing forces

Forces can change the shape of objects.

Elastic objects go back to their old shape when the forces are removed.

Plastic objects don't go back to their old shape.

twisting force	pushing force	bending force	pulling force

♂ Forces can be pushes, pulls, twists, or bends.

Weight

Gravity pulls everything down.

The force of gravity on an object is measured in newtons, with a spring balance.

This force is called **weight**. It is not the same as mass. You will find the weight of a 2 kg mass is very nearly 20 N.

Mass and weight are often confused with each other!

spring balance

NEWTONS

scales

2.00 kg

measures the weight of an object

measures the mass of an object

A spring balance measures the force of gravity on an object: its weight. Scales tell you the mass of an object.

Getting going

Forces can change the motion of an object.

A thrust is a force that pushes an object forward. It can make something start moving, or speed up.

Friction always acts against the direction of motion. It can slow something down.

When the friction force equals the thrust, they cancel each other out. Something that is already moving keeps a steady speed.

Only unbalanced forces can make an object start moving, speed up, or slow down.

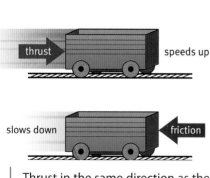

direction of motion

thrust | speeds up

slows down | friction

Thrust in the same direction as the motion makes it go faster. Friction against the motion slows it down.

Rise and fall

Dave kicks a ball into the air.

The ball gets faster as his foot pushes on it.

Then gravity takes over, pulling the ball down. The ball keeps moving upward, but more and more slowly.

The ball reaches its highest point. Now gravity makes the ball fall, faster and faster.

What goes up must come down again.

Drop and drag

Sal jumps out of an aircraft.

She opens the parachute to control her speed.

'The **drag**, or air resistance, gets bigger as I speed up,' she says. 'I get to my top speed when the air resistance exactly balances my weight.'

Sal packs her parachute carefully before each jump. 'If it doesn't open up properly, there won't be enough drag and I'll hit the ground too fast.'

Speed

Radar triggers a speed camera to capture a speeding car.

The camera takes two photos of the car, 0.5 s apart.

The scale on the road shows that the car has moved 9.0 m between photos.

Its **speed** is:

$$\frac{\text{distance}}{\text{time}} = \frac{9.0 \,\text{m}}{0.5 \,\text{s}} = 18 \,\text{m/s}$$

(This is about 40 mph. In a city street, for example, that is too fast!)

Gravity pulls the ball back to the ground.

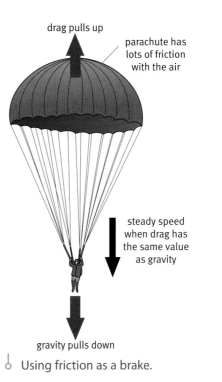

drag pulls up

parachute has lots of friction with the air

steady speed when drag has the same value as gravity

gravity pulls down

Using friction as a brake.

Speed cameras measure distance covered in a set time.

Too much speed can kill.

Speed control

Driving too fast is dangerous. You have less time to react to danger. Even when you do react, the car moves further before it stops. If it hits something, it does so with a greater force. Slower is safer.

There are many ways of forcing cars to stick to safe speeds:
- Install speed cameras and fine reckless drivers.
- Put speed bumps on the road.
- Fit a limiter to the engine, fixing the top speed.
- Set up chicanes to reduce the width of the road.

Calming traffic by forcing it into a single lane from time to time.

Car safety

Cars have lots of safety features built into them. They all work by making the time needed to stop the car or its passengers as large as possible. This reduces the forces on them.
- Crumple zones in the bodywork allow the car to collapse.
- Seatbelts reduce the force needed to stop the passengers.
- Airbags inflate and cushion the impact.

Without these features, passengers would be suddenly stopped by the dashboard and injured by large forces.

Crumpling the car can help protect the passengers.

Into space

Getting into space needs a lot of force. Rockets are the only way. They burn fuel to make hot exhaust gases.

As the exhaust rushes down out of the **rocket**, the spacecraft is pushed up.

A rocket usually has three stages. Each stage in turn pushes the spacecraft up. When each stage has run out of fuel it falls back down to Earth. It may fall into the sea where it can be recovered and used again, or it may burn up in the atmosphere.

Once a spacecraft is in orbit, the rockets are no longer needed.

There is no friction above the atmosphere, so no push is needed to keep going at a steady speed.

Re-entry

Spacecraft move at high speed when they are in orbit.

Most of that speed has to be lost before the spacecraft can land safely.

This is done in three stages:

- The rockets are fired to slow down the spacecraft, making it lose height.
- Friction with the atmosphere slows it down more (and heats it up).
- Parachutes are deployed once the spacecraft is moving slowly enough.

The shuttle glides to land instead of using the parachutes.

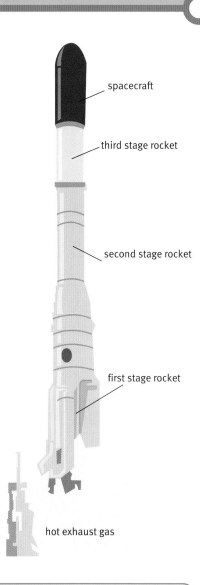

spacecraft

third stage rocket

second stage rocket

first stage rocket

hot exhaust gas

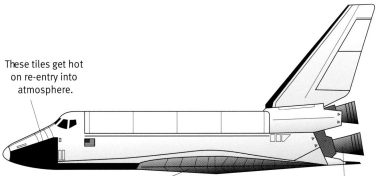

These tiles get hot on re-entry into atmosphere.

Wings allow Shuttle to glide down through air.

The Shuttle is prepared for landing by turning round to orbit tail first. The rocket engines then slow it down.

Questions

1 Write a sentence about each of these words:

elastic   gravity

weight   drag

speed   rocket

2 What do you think of these ideas?

Things can't move up because gravity pulls them down.

Objects falling down just get faster and faster.

You need a source of light to see.

Colour helps you choose the best stuff to eat.

Seeing

Maryam's lamp is **luminous**.

It gives out rays of light when it is switched on.

The rays **reflect** off the walls of the cave.

Some reflected light rays enter her eyes, allowing her to see the cave.

She can't see anything when the lamp is switched off.

Colour

Your eyes collect light from your surroundings.

Your eyes can only detect these primary colours:
- red
- green
- blue.

All the other colours that you can see are called secondary colours.

Each secondary colour contains a different amount of each primary colour:
- Yellow light is a mixture of red and green light.
- Magenta light is a mixture of red and blue light.
- White light is a mixture of red, green, and blue light.
- Cyan light is a mixture of green and blue light.

You only see black when no light is getting into your eyes!

Three spotlights (red, green, and blue) shine their light onto a screen. The secondary colours appear where the circles of primary colours overlap.

All the colours of the rainbow.

ultraviolet   spectrum of visible white light   infrared

Colour temperature

When things get hot enough, they glow. They become luminous.

As things heat up, they start to give off red light. As the temperature rises further the light becomes yellow. At even higher temperatures the light becomes blue.

Astronomers use the colour of light from a star to measure its surface temperature. Our own Sun glows yellow, so it has a surface temperature of about 5000 °C.

Mirrors

Light reflects off anything shiny.

It usually scatters in all directions at random.

The smooth surface of a **mirror** reflects light so that it forms images.

These images appear behind the mirror.

Images in mirrors are just like the real thing – except that left and right are swapped!

Reflections

Shiny surfaces are very smooth.

All light rays coming in at a particular angle reflect off at the same angle.

This doesn't happen on rough surfaces.

Internal reflection

Light can always get into a transparent material.

Sometimes it can't get out again. It is reflected back when it strikes the inner surface at a shallow angle.

A transparent surface can behave like a perfect mirror.

Images in mirrors are slightly different.

An image of distant galaxies, made by light at the end of a long, long journey.

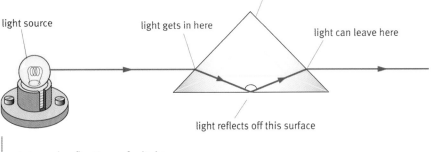

light source

glass block

light gets in here

light can leave here

light reflects off this surface

Internal reflection of a light ray.

Lens image

A **lens** changes the direction of rays of light that pass through it. One type of lens is fatter in the middle than at its edges. This is called a convex lens.

A convex lens collects light from a distant object. It focuses that light to an image on the other side of the lens. It collects light from one point of an object.

The image is not only upside down. It is also smaller than the object.

Images made by lenses are quite different to the real thing.

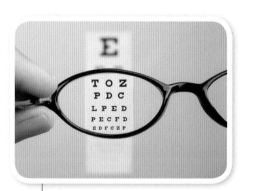
Spectacle lenses for long sight are convex.

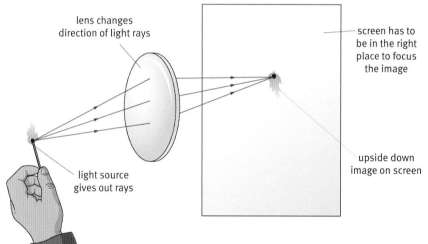
lens changes direction of light rays

screen has to be in the right place to focus the image

light source gives out rays

upside down image on screen

Two convex lenses in a telescope can make far things appear closer.

The convex lens in this projector makes images on a large screen.

Cameras use convex lenses to make images of things.

Optical fibre

Thin strands of pure glass, called **optical fibres**, connect all parts of our world.

Once light gets into these optical fibres, it passes all the way to the other end before it escapes.

The glass is so pure that the light can travel hundreds of kilometres before it is finally absorbed.

The Internet relies on optical fibres to carry huge amounts of data around the world.

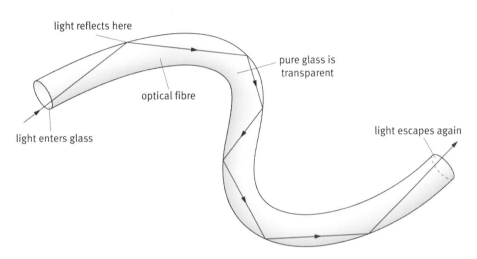

light reflects here

pure glass is transparent

optical fibre

light enters glass

light escapes again

Digital code

The Internet uses optical fibres to carry data very quickly from one computer to another.

The information is coded as pulses of light.

Light can be switched on and off very quickly.

Any information can be coded in this way, including speech, text, and video.

Once the pulses get into the fibre, they are trapped until they get to the other end.

Optical fibres carry more data per second on the Internet than any other technology.

Questions

1 Write a sentence about each of these words:

luminous   reflect   image   mirror   lens   optical fibre

2 What do you think of these ideas?

The Internet works at the speed of light.

You see what your brain thinks you see.

This is your home, photographed from the surface of the Moon.

Earth and Sun

You live on the Earth. It is just a large ball of rock about 13 000 km across.

The Earth moves around the Sun in a circular orbit. The Earth goes once around that orbit each year. That's a journey of almost a billion kilometres at a speed of 30 kilometres per second.

The Earth is not alone. There are eight other **planets** orbiting the Sun, making up the Solar System. The Sun is our nearest **star**. It is much bigger than the planets in orbit around it. The Sun's gravity holds planets in their orbits.

Orbits

Each planet has its own **orbit**, an almost circular path with the Sun at its centre. As the orbit size increases, the orbit time gets longer and the planet moves more slowly.

Planet	Orbit size in millions of km	Orbit time in years	Speed in km/s
Earth	150	1.0	30
Jupiter	778	12	13
Mars	228	1.9	24
Mercury	58	0.2	58
Neptune	4497	165	5.5
Saturn	1427	30	9.5
Uranus	2870	84	6.8
Venus	108	0.6	36

Planets

Eight planets orbit our Sun. Apart from Earth none of them has been visited by astronauts, but all have been studied by space probes.

The planets all orbit the same way around the Sun but at different speeds.

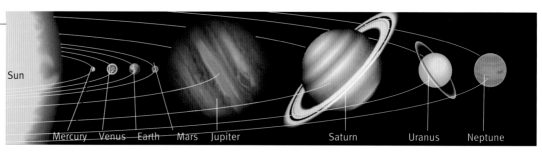

Sun   Mercury Venus Earth Mars Jupiter   Saturn   Uranus   Neptune

Jupiter is almost as big as a star. Mercury is a rocky solid. Saturn is mostly liquid. Venus is close enough to outshine most stars. Uranus is so far away that it can only be seen through a telescope.

Sun

The Sun is a giant fusion reactor. It supplies the energy that makes life on Earth possible.

Deep at its centre, the pressure and temperature are high enough to turn hydrogen into helium.

This reaction transforms (changes) mass into energy at a rate of 4 million tonnes every second. Most of this energy leaves the Sun as visible light.

Planets close to the Sun are hot because they absorb a lot of this light. Faraway planets only absorb a little, so they are cold.

A lot of sunlight leaves the Solar System.

Stargazing

On a clear night, you can see millions of stars. Most stars are like our Sun but so far away that the light we see from them is quite dim.

You can't see any other stars during the day because the Sun is so much brighter. That's because it is our nearest star.

Astronomers use telescopes to investigate stars and planets. They prefer to use sites that are high up on mountains. This has two advantages:
- There is no light pollution from street lights.
- The air is free of clouds and dust, which can block starlight.

Visiting planets

We know a lot about the other planets in the Solar System. This is because we have sent spacecraft to investigate them. None of the spacecraft has been manned. They have been controlled from the Earth. A manned spacecraft needs a lot of things that an unmanned one does not:
- a supply of food, water, and oxygen to keep the astronaut alive
- a source of heat to keep the astronaut warm
- extra fuel to land the astronaut back on Earth at the end of the mission.

Manned spacecraft have only been as far as the Moon.

Looking directly at the Sun is dangerous. Sunglasses reduce the risk of damage to your eyes.

Astronomers prefer to work in dark places where they have a clear view of the night sky. They analyse light from stars to discover how hot they are and what they are made of.

The International Space Station orbits the Earth. The panels of solar cells convert sunlight into electricity.

Satellites

Earth is surrounded by artificial satellites. They are put into orbit using launch rockets. Once they are moving fast enough, Earth's gravity keeps them in a stable orbit. They are above the atmosphere, so there is no air to slow them down.

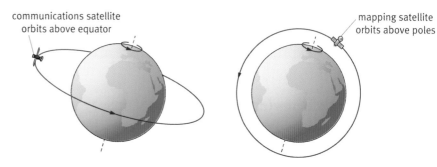

communications satellite orbits above equator

mapping satellite orbits above poles

Using satellites

Communication satellites orbit over the equator. They go around the Earth exactly once every day. This means they remain above the same spot on the Earth.

Mapping satellites orbit over the Earth's poles. They use radar to measure every point on the Earth's surface as it passes underneath.

Spy satellites are in low orbits. Their cameras can see the surface clearly.

GPS tracking satellites constantly radio out their position and the time. People can work out where they are on the Earth's surface by receiving signals from three or more of these satellites.

Moons

Jupiter has 16 moons. This image combines several images to show the relative sizes of four of Jupiter's moons.

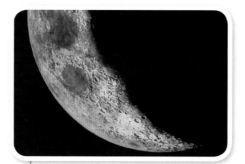

The **Moon** is a natural satellite of the Earth. We can only see the parts that reflect light from the Sun.

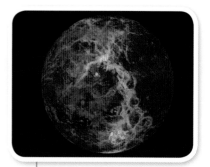

Venus is our nearest planet. It has no moons at all.

Milky Way

Our Sun is one star out of billions in a **galaxy** called the Milky Way. We can see billions of other galaxies with the help of telescopes. Some look just like our own. Others are quite different.

Galaxies may be held together by the gravity of black holes at their centre. Black holes are so massive that not even light can escape from them.

Universe

The stars are not evenly spread across the night sky.

This is the view towards the centre of the Milky Way. Each speck is a star. Our Sun is near the edge.

The **Universe** contains billions of galaxies.

Each galaxy contains billions of stars.

Some of those stars have planets orbiting around them.

At least one planet is inhabited by intelligent life. There may be others.

This image of very distant galaxies was taken by the Hubble Space Telescope.

This is the first image of a planet around another star. It was taken by astronomers in 2004.

Questions

1 Write a sentence about each of these words:

   planet   star   orbit   Moon   galaxy   Universe

2 What do you think of these ideas?

   The Universe is so large that there must be other Solar Systems like ours.

   We only see stars when we can't see the Sun.

Energy

Energy is all around you.

It isn't easy to spot, but life is impossible without it.

For example, the leaves of trees transfer (change) light into chemical energy.

Some of that chemical energy ends up in fruit, which we can eat.

Chopping down the tree transfers its gravitational energy into kinetic energy as it falls.

The chemical energy in the wood can be transformed into heat by burning it.

Many power stations use heat to generate electrical energy.

Trees convert light into chemical stores of energy.

The energy for this aircraft comes from crude oil. Will you still be able to fly when oil runs out?

Making electricity from the wind. We can do this until the Sun stops shining.

Heating homes is the biggest use of energy in the UK.

Renewable energy sources	Non-renewable energy sources
wind	coal
waves	uranium
sunlight	crude oil
hydroelectricity	natural gas

Energy sources

You use lots of energy – about 4000 joules every second. You need it for transport, food, electricity, and heating. Energy comes from two kinds of **energy source:**

- **Renewable** energy sources rely on the Sun to replace the energy we use. The Sun will still be shining in a billion years. Renewable sources will effectively last forever!
- **Non-renewable** energy sources will run out one day. Once they have been used up, there will be no more.

The Earth's population is still rising. All those extra people will want to use energy. Renewable sources will become more important in the future.

Fuels

Petrol (from oil) is a liquid. This makes it easy to transport.

This house was damaged by an explosion after an escape of gas.

Five different **fuels** are used to provide the energy we use.

Only one of them is renewable. We can grow more wood, but oil, gas, coal, and uranium are not replaced as we use them.

No non-renewable fuel is perfect. Each has its own advantages and disadvantages.

Fuel	Energy MJ/kg
gas	55
oil	45
coal	20
wood	16
uranium	500 000

Non-renewable fuels	Gas	Oil	Coal	Uranium
How much CO_2 does it make?	some	quite a lot	lots	little
Is it easy to store?	no	yes	yes	no
How much is there left in the ground?	some	some	lots	lots
Does it burn cleanly?	yes	no	no	doesn't burn
Is its waste radioactive?	no	no	no	yes
How much SO_2 does it make?	little	some	lots	none

Biomass

Almost anything that grows can be used as an energy source. There are two ways of using this biomass:
- Energy released by burning it can be used to boil water and make high-pressure steam. Wood and straw are used to make electricity this way.
- Fermenting with microbes converts biomass into alcohol or methane. Alcohol can replace petrol for cars and methane can replace natural gas in gas turbines.

This biomass is grown for car fuel. Microbes will make alcohol from the sugar in this crop. Alcohol can replace petrol in car engines.

Sea power

There is a lot of energy in waves. Some of that energy can be made into electricity.

Floating tubes on the sea move up and down as the waves pass. This pumps oil through turbines to make electricity.

The tide makes lots of seawater rush in and out of bays twice a day. That water can be trapped behind a special dam called a barrage.

When the water passes through the barrage it spins turbines to make electricity. **Tidal power** works best in places where there is a big difference between high and low tide.

Ground power

Some countries use heat from the ground to make electricity. Cold water is pumped into hot rocks deep underground. The water becomes steam. This goes back up to the surface where it spins turbines to make electricity.

Solar heating

Some people use sunlight to heat up their water. The water trickles over flat metal panels on the roof. The metal is painted black so that it can absorb the sunlight and heat up.

A white surface would reflect the sunlight instead of absorbing it. A transparent protective cover lets the sunlight through to the metal panels.

A Pelamis generator uses **wave power** to produce electricity.

In a geothermal power station energy from hot rocks is used to produce steam to drive turbines.

Water is heated as it passes through these roof panels.

Wind power

Wind turbines change the kinetic energy of the wind into electricity. You need a lot of turbines to get a useful amount of electricity. Here are three ways of increasing the output of a wind turbine:

- Increase the length of its blades so that it catches more wind.
- Hold it higher above the ground where the wind moves faster.
- Keep it facing the direction of the wind.

Wind turbines can only make electricity when the wind blows. They can't provide all of our electricity.

The wind spins the blades round to make electricity.

Hydropower

Hydropower changes the gravitational energy of rivers into electricity. Water at the base of the dam is at high pressure, so can spin a turbine on its way through the dam. Once it gets to the sea, the Sun evaporates it. The vapour cools as it rises, condensing into clouds and eventually falling as rain over the land. That rain means that the dam need never run out of water.

Water flowing through the base of this dam generates electricity.

Solar electricity

Solar cells can transfer light energy directly into direct current electricity. They are expensive, but don't make any pollution. They are becoming cheaper as more people install them. You need a lot of them to replace a single coal-fired power station.

Questions

1 Write a sentence about each of these words:

energy source   renewable   non-renewable   fuel

tidal power   wind turbine   solar cell

2 What do you think of these ideas?

Air travel will be impossible if we run out of oil.

We can get all of our electricity from renewable sources.

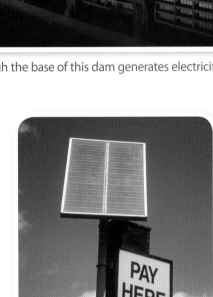
Solar cells are good for providing electricity in remote locations.

The surface of the Moon is covered with craters. What caused them?

This crater is in Arizona, US. The rim is 200 m above the floor.

Craters

A large lump of rock hurtles towards the Moon and hits it.

Its impact is so great that the surface melts. Molten rock is flung up to form the circular rim. The rock left within the rim cools and sets solid to form the flat floor. This is a **crater**.

The surface of the Moon is covered with craters.

There must have been lots of impacts with rocks from space.

Why does the surface of the Earth have so few craters, compared to the Moon?
- Many rocks from space burn up in the atmosphere.
- Craters get worn away by erosion from wind and water.

There is no atmosphere on the Moon, so there is nothing to wear down the craters. The evidence of 4000 million years of impacts is preserved on its surface.

Asteroids

Gravity pulls rocks towards each other. Given enough time, rocks collect together to make planets. Sometimes the gravity of a nearby planet can stop this happening. The asteroid belt between Jupiter and Mars is an example. Jupiter's gravity tears the rocks apart as fast as their own gravity pulls them together. So **asteroids** are rocks left over from long ago, when the Solar System formed.

There are lots of asteroids in orbit around the Sun. Sometimes they collide with each other. This can alter their circular orbits, making them elliptical. The new orbit may take an asteroid close to the Earth.

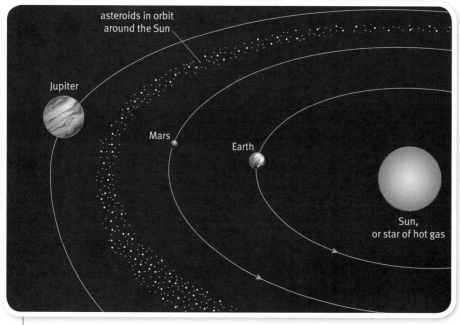

The asteroids between Mars and Jupiter orbit the Sun.

Comets

Comets can also cause serious damage when they hit planets.

A comet is a ball of ice and dust from beyond Neptune.

It has an elliptical orbit, which passes inside the orbit of Mercury.

It moves very slowly when it is far from the Sun.

It speeds up as it falls towards the Sun.

As the Sun heats it up, the comet produces a tail of gas and dust.

Halley's Comet. It takes 76 years to make one orbit of the Sun.

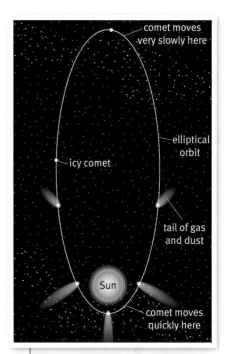

The Sun's gravity makes comets speed up as they get closer.

The Moon

The **Moon** is a large ball of rock that orbits the Earth.

One complete orbit takes 28 days.

The Moon is quite unlike Earth.
- It is too small to keep an atmosphere.
- It has no magnetic field.
- It has no life at all.

However, its rocks are very similar to those found here on Earth.

Making the Moon

Many other planets have moons that orbit around them.

All of these moons are much smaller than their planets.

Our Moon is relatively big. Scientists think it may have formed from a planet that collided with Earth billions of years ago. All of the heavy iron ended up in the Earth, giving it a high density and lots of magnetism. The lighter rocks formed the Moon, giving it a low density and no magnetism.

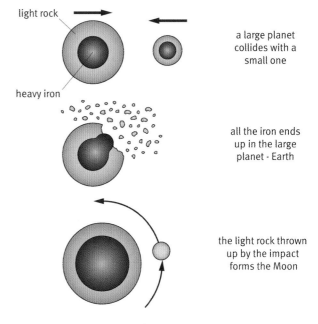

light rock

heavy iron

a large planet collides with a small one

all the iron ends up in the large planet - Earth

the light rock thrown up by the impact forms the Moon

Direct hit

The Earth is about 4000 million years old.

Every 100 million years or so, it is hit by a big asteroid, perhaps 10 km across.

This takes only a few seconds to pass through the Earth's atmosphere. Friction with the air makes it white-hot.

The asteroid slams into the Earth, throwing vast quantities of hot rocks up into space. Gravity quickly pulls them back, setting off fires where they land.

The dust thrown into the air blocks off sunlight, killing many plants. Without plants to eat, many animals die.

The loss of many species of plants and animals at the same time is called a 'mass extinction'. The last mass extinction was 65 million years ago, when the dinosaurs last lived.

An asteroid may have killed off the dinosaurs. Today's scientists do not all agree on this. Some have other ideas about what caused the dinosaur extinction.

Look out!

The Earth moves around the Sun at a speed of 30 kilometres per second through empty space.

Small rocks that get in the way burn up in the atmosphere. Any remains of these meteors that make it to the surface of the Earth are called meteorites. They are usually rich in iron.

Anything in space that crosses the Earth's orbit is called a Near-Earth Object, or **NEO**. Most of them are quite small, but big ones could cause widespread damage if they hit. Astronomers watch carefully for them.

Like planets, NEOs have a different position in the sky each night. Measurements over several nights allow their orbit path to be calculated. A computer works out if the NEO might one day hit the Earth. The nearer an NEO gets, the better its path can be predicted.

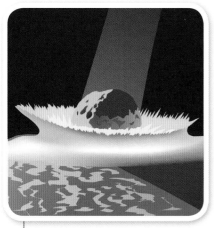

An asteroid collision may have caused the extinction of the dinosaurs.

This car was hit by a meteorite. Luckily, nobody was injured.

Astronomers can only see NEOs in the night sky.

NEO risk

In 2002 an NEO the size of a football pitch passed between the Earth and the Moon. Had it hit the Earth, it might have devastated a whole continent. But it didn't. We had a lucky escape.

So what's the risk from NEOs? Those that are big enough to threaten all life on Earth are rare, perhaps coming along every 1000 million years. Smaller ones cause less damage, but are more frequent.

Size of NEO	Average occurrence
10 m	each year
100 m	every 1000 years
1 km	every 100 thousand years
10 km	every 100 million years

A comet probably caused this damage to Siberia in 1908. The forest was flattened for a hundred kilometres.

NEO action?

The US space agency NASA looks out for NEOs that are over a kilometre across. These are big enough to devastate a whole continent. What could we do if an NEO was spotted heading for a collision with the Earth? It might be possible to:
- use rockets to send a spacecraft to land on the NEO
- use the spacecraft's rockets to nudge the NEO into a different orbit.

Space flight is expensive, so this strategy has not been tested.

Early detection of every NEO and accurate mapping of its orbit may be crucial to our survival. The more time that we have to change an NEO's orbit or prepare for its impact, the better.

Perhaps this technology is what we need to fend off an NEO.

Questions

1 Write a sentence about each of these words:

   crater   asteroid   comet   Moon   NEO

2 What do you think of these ideas?

   Astronomers should be paid for each NEO they discover.

   There is nothing we can do about NEO impacts.

This forecourt pump provides petrol, diesel, and biofuel.

The spark at the tip of the plug sets fire to the fuel.

Liquid fuels

Most vehicles on the road today have engines that burn a liquid fuel. The burning fuel gives out the energy that drives the engine. Four different fuels are used:

- petrol
- diesel
- liquid petroleum gas (LPG)
- biofuel.

The first three come from crude oil. The last one comes from fermenting plants or processing plant oils.

Four-stroke engine

Most transport engines transfer energy from their fuel in four steps:

- A piston draws fresh air into a cylinder.
- The piston compresses the air, raising its temperature and pressure.
- Fuel is sprayed into the cylinder and ignited; the gases in the cylinder get very hot and push the piston down the cylinder.
- A valve opens to let the exhaust gases escape into the air as the piston goes up again.

Then the cycle starts again.

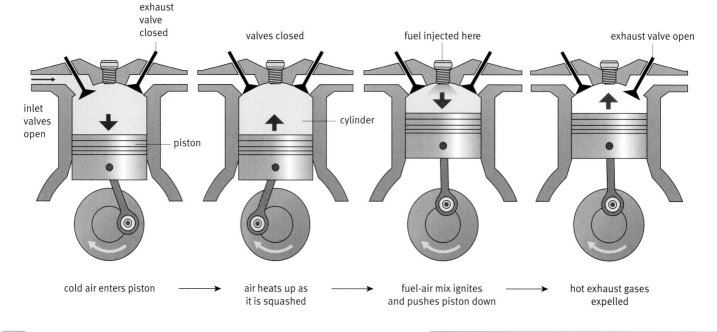

cold air enters piston → air heats up as it is squashed → fuel-air mix ignites and pushes piston down → hot exhaust gases expelled

Ignition

Diesel fuel ignites on its own when it is injected into the cylinder. The compressed air is hot enough to do this. Other fuels need a higher temperature for ignition. This comes from a spark plug. A quick high-voltage pulse of electricity sets the air–fuel mixture on fire at the right time.

Less effort

Jim fits new tyres onto cars. "I use a jack to raise the tyre off the ground," he says. "I'm not strong enough to lift a car by myself!"

Jim uses a spanner with a long handle to undo the nuts holding the wheels in place. "I can't undo the nut with my bare hands," says Jim. "The long handle allows me to do the same job using less force."

Jacks and spanners are examples of machines. They both allow Jim to convert his small effort force into a larger load force.

This jack converts Jim's sideways effort force of only 100 N into an upwards load force of 10 000 N on the car.

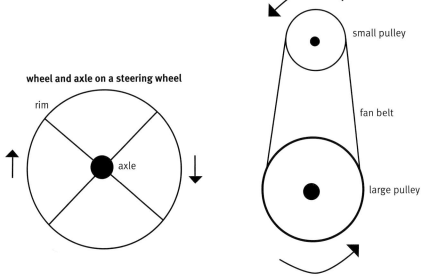

wheel and axle on a steering wheel

belt and pulleys for the engine-cooling fan

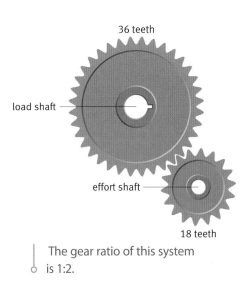

The gear ratio of this system is 1:2.

Many machines use gears.

The ratio of the number of teeth on the gears is the ratio of the effort force to the load force.

The gear wheel on the effort shaft has 18 teeth. The gear wheel on the load shaft has 36 teeth.

The gear ratio is 18:36 or 1:2. The force from the load shaft is twice as big as the force on the effort shaft.

Electric motors

Electric motors have lots of uses in a car. Two uses are very important:

- starting the engine
- moving the wipers across the windscreen when it rains.

Both motors work the same way. Electric current from the battery passes through a coil of wire. The current interacts with a magnet to spin the coil.

Windscreen wipers increase your driving safety.

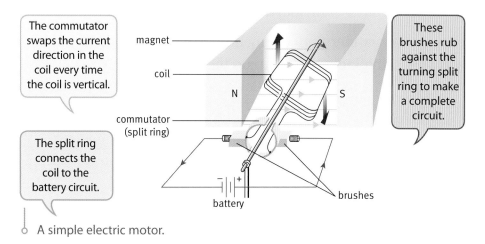

The commutator swaps the current direction in the coil every time the coil is vertical.

The split ring connects the coil to the battery circuit.

These brushes rub against the turning split ring to make a complete circuit.

magnet

coil

commutator (split ring)

battery

brushes

A simple electric motor.

Electric cars

Some cars only have electric motors. They store their energy in large batteries. The batteries have to be charged up from the mains supply at the end of a day's driving. There are many reasons why solar cells on the car roof aren't used instead:

- Solar cells don't work in the dark and the Sun doesn't always shine during the day.
- They cost a lot of money to install.
- They can't provide the car with enough energy to go fast.

Both motors work the same way. Electric current from the battery passes through a coil of wire. The current interacts with a magnet to spin the coil.

Charging up an electric car in the street.

Speed

Suppose a car takes 10 minutes to cover a distance of 6 km. How fast is it moving? You need to calculate its speed in metres per second.

6 km is 6000 metres and 10 minutes is 600 seconds so

$$\text{speed} = \frac{\text{distance}}{\text{time}} = \frac{6000 \text{ m}}{600 \text{ s}} = 10 \text{ m/s}$$

This is about 20 mph, a safe speed near a school.

Maximum speed

There are two reasons why every road has a speed limit:
- It reduces your chance of having an accident.
- It prevents you from wasting fuel by going too quickly.

The national speed limit depends on the type of road:
- 70 mph on a motorway
- 60 mph on most other roads.

Some roads have lower speed limits. This protects cyclists and pedestrians:
- 30 mph in towns
- 20 mph near schools.

Speed bumps and speed cameras encourage people to keep to these limits.

Stopping

The stopping distance of a car is the sum of two separate distances:
- The thinking distance – people take about half a second to notice danger and get their foot on the brake. The car is still going at full speed during this time.
- The braking distance – the brakes take time to slow the car down. During that time the car keeps moving forward, but less and less quickly.

Questions

1 Write a sentence about each of these words:

fuel   petrol engine   machine   gear ratio
electric motor   speed   stopping distance

2 What do you think of these ideas?

Only electric cars should be allowed in town centres.

Driving would be cheaper if the national speed limit was 50 mph.

This sign warns you to limit your speed.

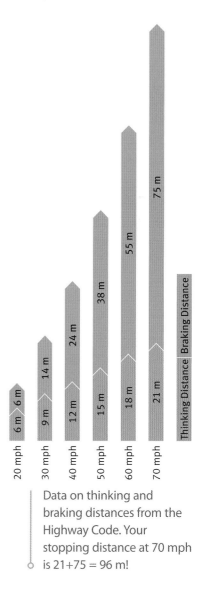

Data on thinking and braking distances from the Highway Code. Your stopping distance at 70 mph is 21+75 = 96 m!

Anita is refuelling. The chemical energy stored in her food came from the Sun as light.

Energy flows through the phone while it is being used.

Energy from the warm liquid transfers to the cold ice.

Energy

Energy is everywhere.

For example, light transfers energy from the Sun to the Earth.

Some of this is stored by plants as chemical energy.

After you eat the plants, you change their chemical energy into other types of energy:

- kinetic energy as you move
- **heat** to keep you warm
- sound as you talk and sing.

Energy transfers

Energy is only really useful when it is on the move.

Jack's phone only comes alive when he turns it on.

The battery is a chemical store of energy.

That energy is used in several different ways:

- Microwave signals carry energy to the nearest base station.
- Sound waves transfer energy to Jack's ears.
- Light transfers energy from the display to Jack's eyes.
- Energy is transferred to the surroundings as the phone heats up.

Heat transfers

You are losing energy all the time. This is because your body is warm and its surroundings are cooler. You wear clothes to reduce this energy transfer. Some heat transfers are useful:

- Power stations use heat from burning fuels to produce steam that turns turbines.
- Ovens use heat to cook food.

Heating houses

Sam turns on the heating when she gets home.

'The **temperature** goes up quickly at first,' she says. 'Then it levels off.'

After about an hour, energy escapes from the house as fast as her heating puts it in. The temperature stabilises.

Sam turns off the heating when she leaves in the morning. By the time she gets back, all of the energy she put into her house has transferred to heating the air outdoors. The house is cold again!

The red spots show where most of the energy is escaping from the house.

Heat and power

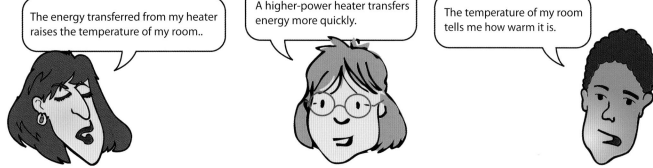

The energy transferred from my heater raises the temperature of my room..

A higher-power heater transfers energy more quickly.

The temperature of my room tells me how warm it is.

Mass, heat, and temperature

Jo is a cook. She needs to know about the difference between heat and temperature.

A large pan of water takes longer to boil than a small one. 'Both eventually get to a temperature of 100°C,' says Jo. 'But the large pan needs a lot more heat.'

The temperature rise of an object being heated depends on many things:
* its mass
* the material of which it is made
* the amount of energy transferred to it.

The large bits of food need more energy to get hot than the small ones.

Quantity	power	heat energy	temperature
Unit of measurement	W	J	°C

Ice to steam

Ice is solid water. It melts at 0 °C. Melting involves a change in the arrangement of molecules. All of the molecules gain energy. If you heat ice, its temperature won't go above 0 °C until it has all melted.

When water reaches 100 °C it can turn into a gas. Boiling involves another change in the arrangement of the molecules. Again, all of the molecules gain energy. When you heat water, it stays at 100°C until it has all changed into a gas.

The reverse processes are condensation and freezing. These release energy as the molecules are rearranged.

This mirror in the desert focuses sunlight onto a metal tube. The water in the tube boils to steam. The steam spins a turbine to make electricity.

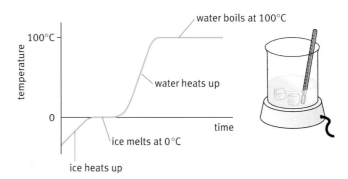

Solar cooking

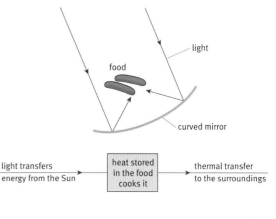

Cooking with sunlight.

Hot air

You lose a lot of energy without clothes on.

This is because air expands as it is heated, making it lighter. It then floats upwards, to be replaced by heavier cooler air.

Your clothes trap warm air.

Hair also traps air. It is good at stopping warm air from moving away. The hot air in a balloon makes it rise up through the cooler air around it.

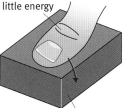

Conductor, insulator

Metals feel cold when you touch them. This is because energy flows easily through them when there is a temperature difference.

As the energy flows away from your skin, its temperature drops. It feels cold. Metals are good **conductors**.

Insulators feel warm to the touch. Energy flows slowly through them, leaving the temperature of your skin unchanged. Wood and plastic are insulators. So are materials that contain lots of pockets of air.

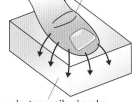

finger feels cold because it loses energy

finger feels warm because it loses little energy

conductor easily absorbs energy from the warmer finger

insulator absorbs very little energy from the warmer finger

The air bubbles trapped in the plastic slow down the transfer of energy from the hot drink.

Hot kitchen

Good kitchen equipment uses conductors and insulators carefully. Pots and pans made of metal allow food to heat up quickly. Handles made of wood or plastic stop people getting burnt. Fluffy material in the walls of fridges stops heat getting in easily.

Home insulation

Heating houses in the UK costs about £10 billion a year. A lot of money can be saved by putting in insulation. The insulation pays for itself through the money saved on heating bills. How long this takes is called the payback time.

Frying pans get hot.

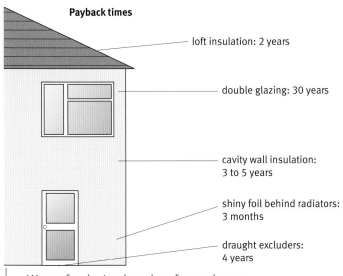

Payback times

loft insulation: 2 years

double glazing: 30 years

cavity wall insulation: 3 to 5 years

shiny foil behind radiators: 3 months

draught excluders: 4 years

Ways of reducing heat loss from a house.

Questions

1 Write a sentence about each of these words:

heat   temperature

conductor   insulator

2 What do you think of these ideas?

Adding energy to objects always makes them hotter.

People should be forced to insulate their houses.

Making electricity

Most of your electricity is made by moving magnets.

The magnet needs to be near a coil of wire.

A voltage is produced in the coil each time the magnet is moved. There is no voltage when the magnet stays still.

There are several ways of increasing the voltage:
- use a stronger magnet
- move it more quickly
- have more turns on the coil
- wind the coil around an iron core.

Moving a magnet towards a coil of wire makes a current flow in one direction. Moving it the other way makes the current flow in the other direction.

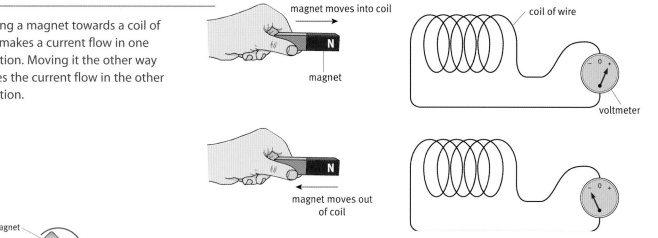

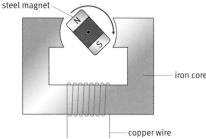

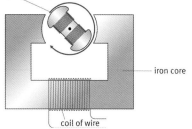

Generators

An electricity **generator** has three parts:
- a magnet that is rotated
- a coil of wire where the electricity is produced
- a core of iron to carry the magnetism from the magnet to the coil.

A power station generator makes alternating current. It needs to be big to make lots of electricity. So it spins electromagnets instead of magnets.

In the UK, the electromagnets spin round 50 times a second to generate 230 V at 50 Hz.

Using uranium

About 20% of our electricity is made in nuclear power stations. They use **uranium** as their energy source. This is a non-renewable energy source, so it will not last for ever.

Splitting uranium atoms releases a lot of energy. This energy is used to make high-pressure steam. The steam spins a turbine attached to a generator.

Some people would like us to use more nuclear power. This is because waste from uranium fuel is not a cause of global warming.

Special clothing stops radioactive material getting inside the body.

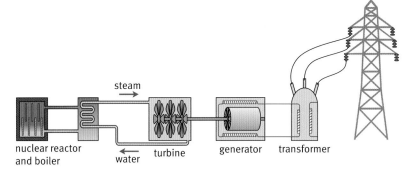

nuclear reactor and boiler | water | turbine | generator | transformer | steam

Radioactive waste

Radioactive waste from nuclear power stations is dangerous. It gives off nuclear radiation.

Nuclear radiation is waves or particles that can reach the cells of your body. Sometimes the radiation starts cancer.

People who work with radioactive materials need to be careful. They must:
• keep their distance
• limit their exposure time
• label places where the waste is stored
• wear protective clothing.

Alpha, beta, gamma

Radioactive materials emit three different types of radiation.

Each needs a different amount of stuff to stop it:
• Alpha particles are stopped by a sheet of paper.
• Beta particles are stopped by a few millimetres of metal.
• Gamma rays are only stopped by several centimetres of lead.

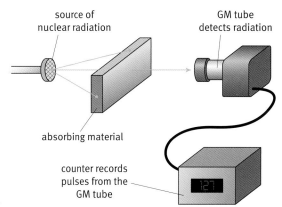

source of nuclear radiation

GM tube detects radiation

absorbing material

counter records pulses from the GM tube

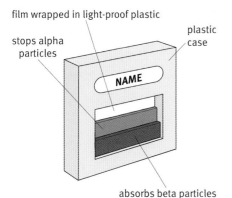

film wrapped in light-proof plastic

stops alpha particles

plastic case

NAME

absorbs beta particles

Henry wears a film badge at work.

Worker protection

Henry works in a nuclear power station.

'I have to put on a film badge as soon as I get to work,' says Henry. 'It records the radiation I am exposed to during the day.'

The film in Henry's badge is routinely processed every eight weeks. Its colour shows how much radiation he has absorbed. It should only be a tiny amount.

If Henry is doing something that could give him a higher exposure than usual, his badge will be processed more often. If he is exposed to too much radiation, he must stay at home for a while – on full pay!

Plutonium

Nuclear power stations make **plutonium**. It is one of the waste products left over from splitting uranium atoms. Plutonium is radioactive. It is also the raw material for nuclear bombs.

All of the plutonium on Earth has been made in nuclear reactors. As soon as it has been made, it starts to decay. Over a few thousand years it turns into stable stuff that is not radioactive.

Radioactive decay

Radioactive materials have unstable atoms.

As each atom decays, it spits out nuclear radiation, becoming a different, stable atom.

Each atom can only decay once, so all radioactive materials become harmless in time.

Some become harmless in a few hours. Uranium stays radioactive for thousands of millions of years.

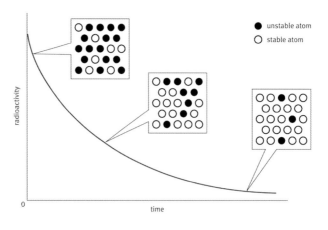

● unstable atom
○ stable atom

radioactivity

time

0

Waste disposal

Nuclear power stations have a problem. They need to dispose of their waste carefully.

Some of it will be radioactive for several thousands of years.

What happens to nuclear power stations at the end of their life?

There are plans to dismantle them with robots. That will be very expensive. At the moment, old nuclear reactors are just kept under guard.

Nuclear safety

Government agencies keep an eye on the nuclear power industry, to ensure that nuclear power is safe.

They do safety tests on a power station and decide when it has come to the end of its useful life.

They also:

- make sure that none of the plutonium is stolen
- check any plans for new power stations
- check that the disposal of radioactive waste is done properly.

fuel rods are removed from the reactor

↓

they are left in water until radioactivity has died down a bit

↓

the fuel is processed → plutonium is extracted

↓

radioactive waste is packed into drums

↓

the waste is stored in a secure location

The uranium is packed into metal tubes before it goes in the reactor. The radioactive waste stays in the tube, so it can be easily removed.

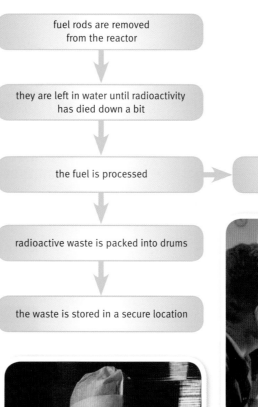

Nuclear power stations are inspected regularly.

Questions

1 Write a sentence about each of these words:

generator   uranium   radioactive   plutonium

nuclear safety

2 What do you think of these ideas?

There is no really safe way of storing radioactive waste.

Nuclear power is a good way of making electricity.

Not everyone is convinced that nuclear power is safe.

A rainbow shows the spectrum of colours in sunlight.

Rainbow colours

There are many ways of separating white light from the Sun into its different colours:

- a rainbow produced naturally by water droplets in the sky
- transmission through a glass prism
- reflection from thin layers of liquid
- reflection from the surface of a CD.

You can remember the order of the colours in the **spectrum** with a sentence like this:

Richard Of York Gave Battle In Vain

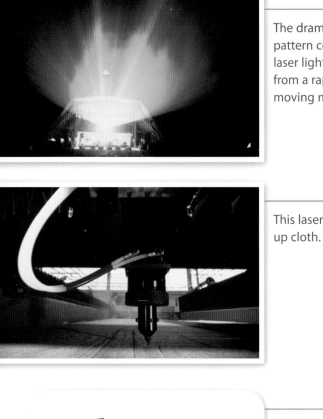

The dramatic pattern comes from laser light reflected from a rapidly moving mirror.

This laser is cutting up cloth.

A laser makes a narrow beam of light.

Laser

Most light sources give out light in all directions, with a range of colours. A **laser** is different:

- The light has only one colour.
- It can transfer a lot of energy.
- All the light goes in the same direction.
- The light can be focused down to a tiny spot.

These properties make lasers ideal for reading information from a CD.

Lasers have lots of other uses as well.

This bomb glides to hit a laser spot shone onto its target.

Electromagnetic spectrum

Visible light is an electromagnetic wave.

It transfers energy through empty space at the amazing speed of 300 000 km per second.

Each colour in visible light has a different wavelength.

Violet has the shortest wavelength. Red has the longest.

Infrared has a wavelength longer than red. This means the human eye cannot see it.

The electromagnetic spectrum. Shorter wavelengths can carry more energy.

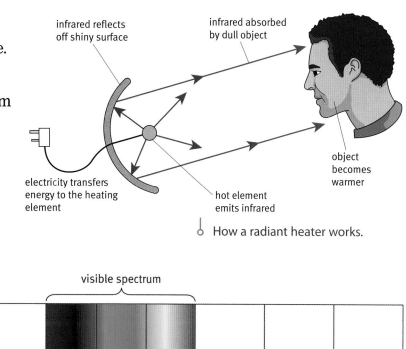

infrared reflects off shiny surface

infrared absorbed by dull object

electricity transfers energy to the heating element

hot element emits infrared

object becomes warmer

How a radiant heater works.

visible spectrum

| UV | violet | green | red | infrared | microwaves | radiowaves |

increasing wavelength

Infrared heating

Infrared waves transfer energy from hotter things to cooler ones.

TV remote

Pulses of infrared make useful links between mobile phones or computers. They can transfer data quickly.

Data transfer is error free when the pulses at the receiver are big enough.

Infrared beams spread out as they travel, so their range is limited to a few metres.

Best of all, you cannot hear or see them, so they are not annoying!

This remote control uses pulses of infrared to transfer information to the TV.

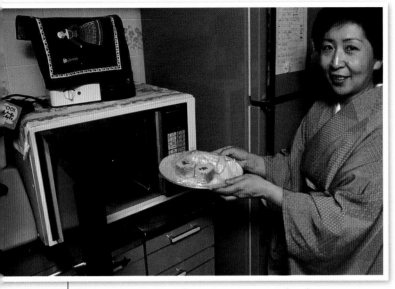

You contain a lot of water, so you need to keep away from microwaves.

This image was made from infrared rather than visible light. Hotter areas emit more infrared than cooler areas.

Night sight

Images made with infrared show differences in temperature.

Warm things stand out against cool ones.

This is because increasing the temperature of an object increases the amount of **infrared** emitted.

Cameras sensitive to infrared allow you to see in the dark.

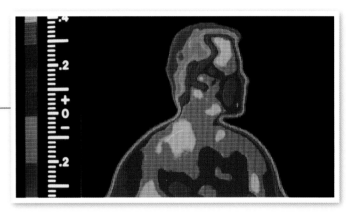

Microwave cooking

Microwave ovens are good at heating food quickly.

This is because they transfer their energy deep inside the food.

But you have to be careful.

Only the water and fat in the food absorbs the microwaves, gaining energy.

Dry things, such as the food container, don't heat up.

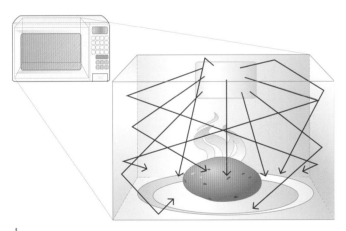

Microwaves in the oven reflect off the metal walls. They are transmitted through the ceramic plate. Water in the potato absorbs the microwaves, getting hotter.

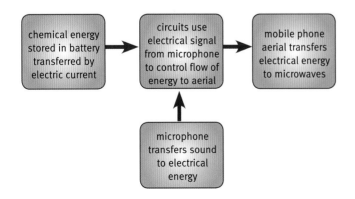

Mobile phone

Mobile phones use pulses of microwaves to communicate with the nearest phone mast.

The power of the microwaves is very low. They can only heat your brain by a tiny amount.

Are mobile phones risky?

How can you find out whether a mobile phone might cause brain cancer?

You can't experiment directly on people. You have to let people experiment on themselves.

Some people use mobile phones. Others don't. So scientists record how many people in each group get brain cancer.

This takes years and years. So far, the results suggest that the risk from mobile phones is very small.

Are they cooking their brains?

Radio waves

Radio waves have a longer wavelength than microwaves.

This means they carry even less energy. They are quite harmless.

Texting may be safer as it keeps the phone away from your head.

Radio, mobile phones, and some laptop computers use wireless links.

Wireless links

Pete is a policeman. He uses a radio link to keep in contact with headquarters.

It isn't very secure because other people can pick up the waves from his aerial.

'Police have their own wavelengths,' says Pete. 'It's illegal to listen in to our messages.'

Pete's radio only works over short distances. As the waves move away from him, they spread out and get weaker.

Questions

1 Write a sentence about each of these words:

   spectrum   laser

   infrared   microwave

   radio wave   wireless

2 What do you think of these ideas?

   Mobile phones should be banned until proved to be safe.

   Microwave ovens are safer than gas ovens.

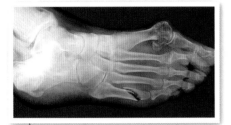

X-rays allow doctors to see inside a patient's body.

Doctor, doctor

Mark falls over and hurts his foot.

He goes to the hospital. The doctors aren't sure if he has torn a ligament or broken a bone.

They study an X-ray image of his foot. This confirms their **diagnosis** of a broken bone.

The medical **treatment** is to put his foot into a plaster cast for a few weeks.

Finding out

Your arm hurts. How does your doctor find out what's wrong?
- She can use her fingers to feel your arm.
- She can use X-rays to make an image of the arm.
- She can give you an anaesthetic and cut the arm open.

The surgical procedure has some risks. Microbes might get in through the cut skin and give you an infection. You might not come round from the anaesthetic.

Exposure to X-rays slightly increases your risk of getting cancer some time in the future. But imaging with X-rays is quicker and safer than surgery.

source of X-rays

bone absorbs X-rays

flesh lets X-rays through

film wrapped in lightproof jacket records arrival of X-rays

Making an X-ray image. The white bits are shadows where the X-rays have been absorbed by bone or metal.

X-rays

X-rays are part of the electromagnetic spectrum.

They pass through skin, muscle, and fat with hardly any transfer of energy.

But X-rays are absorbed by bone or metal. So they are good at making images of bones.

People who work with X-rays take precautions to reduce their risk of cancer:
- They don't go close to X-ray sources.
- They wear film badges to measure their exposure.
- They use metal shielding when the X-rays are switched on.

Risk analysis

Your risk of cancer increases slightly every time you are exposed to X-rays.

The risk from rotten teeth or broken bones is far greater. It makes sense to use X-rays to diagnose these conditions, so that they can be treated.

Bone scan

Tracey has a bone scan.

The nurse injects her with a radioactive liquid.

She lies down in a machine that uses a gamma camera. The camera detects gamma rays from the radioactive liquid as it moves around her body.

A computer uses information from the camera to make an image.

The image shows where Tracey's blood is absorbed into growing bone.

Gamma rays

Gamma rays are like X-rays, but carry more energy.

Most gamma rays pass straight through flesh, skin, and bone.

The few that are absorbed can sometimes make cells turn cancerous.

You need a lot of thick lead shielding to protect yourself from gamma rays.

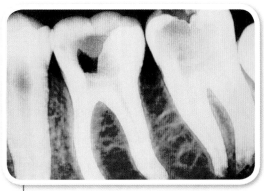
This X-ray image of teeth shows early signs of decay. Prompt treatment will save the tooth.

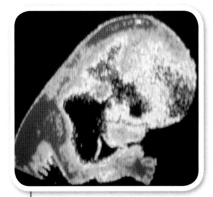
Ultrasound was used to make this image of an unborn child. X-rays were not used. The risk to the baby was too great.

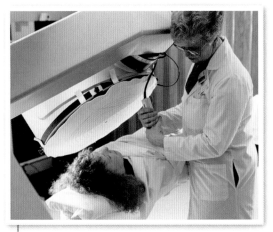
This equipment detects gamma rays from the patient's body.

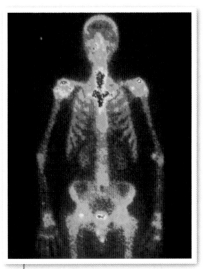
The image shows that bone is growing most in red areas.

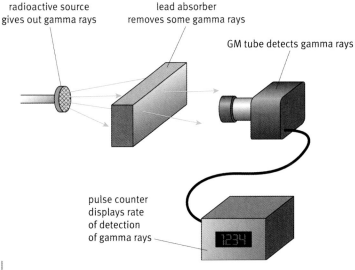

radioactive source gives out gamma rays

lead absorber removes some gamma rays

GM tube detects gamma rays

pulse counter displays rate of detection of gamma rays

Detecting gamma rays in a laboratory.

Fair skin is good at using UV from sunlight to make vitamin D.

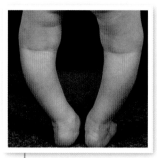

Your bones can go soft if you don't have enough vitamin D.

UV therapy

Ultraviolet (UV) waves are part of the electromagnetic spectrum.

Its wavelength is a bit shorter than violet light, so UV is invisible.

It has some medical uses:
- hardening white tooth fillings
- treating eczema.

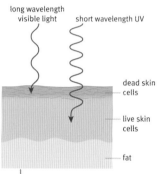

long wavelength visible light

short wavelength UV

dead skin cells

live skin cells

fat

UV can get to the (live) cells just under the skin.

A suntan gives some protection from UV in sunlight.

Suntan

Exposure to UV increases your risk of skin cancer.

Your body uses UV to make vitamin D, but too much UV gives you sunburn.

The pain of sunburn tells you that damage is being done.

Your body responds slowly by putting brown pigment into your skin. The brown pigment absorbs UV and reduces the risk of skin cancer.

Sitting in the shade protects you from UV damage. Clothing helps too!

The SPF number tells you how many times longer it will take for your skin to redden, compared to the reddening time without suncream.

Sunscreen

Jane and Nigel go to the beach to sunbathe.

Nigel doesn't use suncream. His skin burns after only an hour.

Jane has more sensitive skin. Without suncream she burns in 15 minutes. Using suncream with a *sun protection factor* (SPF) of eight means she does not burn for more than two hours.

Electromagnetic spectrum

The penetration of an electromagnetic wave depends on its wavelength.

Short wavelength waves go deeper than long ones.

So visible light is absorbed at the surface of your skin and gamma rays go straight through you.

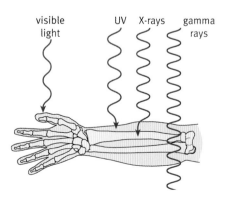

visible light   UV   X-rays   gamma rays

Background radiation

Radioactivity is all around you.

Most of it comes from natural materials.

These include the ground, the air, your food, and your drink.

Cosmic rays come from space.

Only a small amount comes from artificial sources, such as X-rays and nuclear power stations.

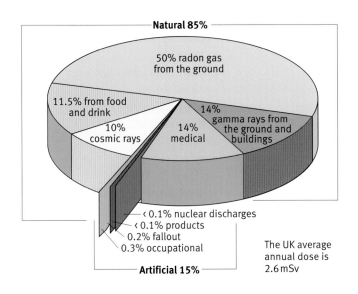

Natural 85%

50% radon gas from the ground

11.5% from food and drink

10% cosmic rays

14% medical

14% gamma rays from the ground and buildings

< 0.1% nuclear discharges
< 0.1% products
0.2% fallout
0.3% occupational

Artificial 15%

The UK average annual dose is 2.6 mSv

Radiation dose

Your risk of getting cancer from radiation depends on your **radiation dose.**

A dose of 1 millisieverts (mSv) gives you a 3 in 100 000 risk of getting cancer. This means that three people in a hundred thousand will get cancer from 1 mSv of radioactivity.

Your radiation dose from **background radiation** is about 3 mSv each year.

Source of radiation	Radiation dose in mSv	Cancer risk
medical X-ray	0.03	1 in 1 000 000
aeroplane flight	0.3	1 in 100 000
background	3	1 in 10 000

Questions

1 Write a sentence about each of these words:

   diagnosis   X-ray

   treatment

   gamma ray

   ultraviolet (UV)

   radiation dose

   background radiation

2 What do you think of these ideas?

   Anything that can give you cancer should be banned.

   All medical treatments increase your risk of damage.

Index

OXFORD
UNIVERSITY PRESS

Great Clarendon Street, Oxford OX2 6DP

Oxford University Press is a department of the University of Oxford.
It furthers the University's objective of excellence in research,
scholarship, and education by publishing worldwide in

Oxford New York

Auckland Cape Town Dar es Salaam Hong Kong Karachi
Kuala Lumpur Madrid Melbourne Mexico City Nairobi
New Delhi Shanghai Taipei Toronto

With offices in

Argentina Austria Brazil Chile Czech Republic France Greece
Guatemala Hungary Italy Japan Poland Portugal Singapore
South Korea Switzerland Thailand Turkey Ukraine Vietnam

Oxford is a registered trade mark of Oxford University Press
in the UK and in certain other countries.

British Library Cataloguing in Publication Data.

Data available.

ISBN 978-0-19-913852-4

10 9 8 7 6 5 4 3 2 1

Printed in Singapore by KHL Printing Company.

Paper used in the production of this book is a natural, recyclable product
made from wood grown in sustainable forests. The manufacturing process
conforms to the environmental regulations of the country of origin.

Acknowledgements

Illustrations by Kamae Design, Mark Walker, and Q2A Media.

Although we have made every effort to trace and contact all copyright holders
before publication this has not been possible in all cases. If notified, the publisher
will rectify any errors or omissions at the earliest opportunity.

Project Team acknowledgments

These resources have been developed to support teachers and students
undertaking the OCR GCSE Science Twenty First Century Science suite of
specifications. They have been developed from the 2006 edition of the resources.

We would like to thank David Curnow and Alistair Moore and the examining
team at OCR, who produced the specifications for the Twenty First Century
Science course.

Authors and editors of the first edition

We thank the authors and editors of the first edition, Jenifer Burden,
Peter Campbell, Donna Evans, and Andrew Hunt.

Many people from schools, colleges, universities, industry, and the professions
contributed to the production of the first edition of these resources. We also
acknowledge the invaluable contribution of the teachers and students in the Pilot
Centres.

The first edition of Twenty First Century Science was developed with support from
the Nuffield Foundation, The Salters Institute, and The Wellcome Trust.

A full list of contributors can be found in the Teacher and Technician resources.

The continued development of Twenty First Century Science is made possible by
generous support from:

- The Nuffield Foundation
- The Salters' Institute